W9-CXY-002

Geographical Literacy

Geographical Literacy

Kieran O'Mahony

Educare *Press*   *Seattle, WA.*

PUBLISHED BY
Educare Press
PO Box 31511, Seattle, Washington 98103

Copyright © 1992 by Kieran O'Mahony

All rights reserved. No part of the contents of this book may be reproduced or transmitted in any form or by any means without the written permission of the publisher.

Library of Congress Cataloging-in-Publication Data: 92-70099
O'Mahony, Timothy Kieran.
    Geographical Literacy/ Kieran O'Mahony
    What Every American Should Know About Geography...and More.
    Includes Index.

ISBN 0-944638-06-6 : $14.95

Printed and bound in the United States of America.

1 2 3 4 5 6 7 8 9
Back cover photograph:
Ronan's "global glow" by Johnny Sheehan.

To
Bernadette,
Shane, and Ronan.

Contents

Other titles from **Educare Press**

Geography and Education
Through the Souls of Our Feet
Kieran O'Mahony

The Valiant Captains
Epics of the Sea - Historical
Sheldon A. Jacobson

The Man Who Moved the World
Archimedes - Historical
Sheldon A. Jacobson

Fleet Surgeon to Pharaoh
Ancient Egypt - Historical
Sheldon A. Jacobson

To The Woods and Waters Wild
Short Stories
Ross Brinn

Acknowledgments

In an undertaking of this magnitude there are countless people and organizations that I need to mention and thank. I am grateful for the encouragement and help selflessly offered by my friends and acquaintances over the past number of years. I particularly wish to thank the Sunday Night Session for putting up with the tuning. Thanks also to Lynn for a Zeutonhorst of a different color, to Gary for the mathematical equations and the trips home. I used to think that Dodge and Bentley were just old cars, now I know they're old friends. To Steve Nieker who recently transmuted from a simple human being to a serious sailor under geographical instruction in the Puget Sound. To Brooks and Creegan for the wonderful sojourn in the Crescent. To Mary and Dennis for Miltown. To Fernando for his dedicated environmental endeavors in Paraguay; I wish him great success. To Jukes for the super-imposed paradigm at Cappoquin, and to JK for the flat-bottomed boat trip down to Mallow. Thanks too, to Bridie and Jack at Shanaway, to all the folks at Rathmore and back at the Castle.

We must all share in a special thanks to the National Geographic Society for its undaunting promotion of things geographical in the last one hundred years and more. I am particularly grateful for the help I received from the Geography Education Foundation in preparing this material. A special thanks also to the Association of American Geographers for their dedicated endeavor and to the many enthusiastic individuals who contribute daily to Alliance courses and fieldtrips.

List of Figures

Preface

Ptolemy drew a map of the world nearly two thousand years ago, a map which has a profound influence on you and me, today. It was not an accurate map, since there were many uncharted places, where man had not yet been. His map represented the known world, and beyond that was a large grey area which he called *Terra Incognita*, or unknown land. He calculated the missing parts of the earth, as best he could, but they were gravely faulty. He exaggerated the size of the land mass that extended from Spain to China and he underestimated the reach of the intervening ocean. This miscalculation had incredible consequences, because Ptolemy's map encouraged a young adventurer to undertake a journey into *Terra Incognita* in the year 1492.

The man was, of course, Christopher Columbus, and the rest is ... geography.

When we look back at our ancestors and see what they knew, and how they managed their lives, we gain a great deal of perspective. We realize that they knew very little about our planet earth, and what they thought they knew, was wrong. But this fact begs the question that you and I must ask of ourselves. What about our descendants? Will they look back at the last few years of the twentieth century and say the same thing about us? Is this our legacy?

Will they say that what we knew about our planet Earth was not much and most of it was wrong? Will they say that we were the generation that punctured the ozone layer, depleted the oxygen supply, denuded the forests, poisoned the water and built gigantic concrete jungles? Or will there be anyone left to look back?

Will some other civilization stumble upon our smoulder-
ing planet and infer that we were the ones who defeated nature,
became technically competent and then self-destructed?

It is time to ask some fundamental questions. Who are
we? What do we want? What of our children? What about our
Planet?

The need for a national commitment to geographical
literacy has never been greater. Despite a record level of public
spending on education, there remains an unacceptably high level
of geographical illiteracy among our young people. Youngsters
lacking the basic skills of geographic literacy have difficulty
interpreting their world as they are confronted by serious challenges in the work-place, especially in this information age. And
the prognosis towards the year 2,000 is even worse.

The message from industry leaders and governmental
bodies is loud and clear. America is not prepared for the technologically advanced society that we have rushed into and
where we now find ourselves. We are ill-prepared for the challenges of the twenty-first century. We have accepted the comfort
and security spawned of today's technological age, but around
us we wreak havoc on ourselves, on our fellow men and especially on our planet.

It is my purpose and the aim of this book to help you and
your children take the first steps to gaining geographical literacy,
and becoming knowledgeable and aware, so that you are able to
make informed decisions about your planet and your lives.

Kieran O'Mahony

If a nation expects to be ignorant and free in a state of civilization, it expects what never was and never will be.

Thomas Jefferson

Chapter One:

On the ground

We are a nation of illiterates, if we go by some people's statistics, at any rate. And yet we're not doing too badly, are we? Some of us can't read, some more of us can't write; functional illiterates they call us. We find it difficult to fill out simple forms, it seems. Simple! Sure ... maybe, simple to some people! Most of us don't ever want to sit in front of a computer, and now they tell us that we are behind in our knowledge of geography. Behind? Behind who? Whom? Geography! What is that? It didn't seem important before! So why all the commotion about it now? Who comes up with these figures and statistics anyway?

Ah yes! Geography, where would we be without it? I'm sure you wake up in the morning with this question on your mind. And if not, maybe you should.

"Where is my geographer, I can never find him when I need him?"

In truth, you don't often hear people ask this question. Geographers are not like your everyday doctor, your neighborhood policeman or your friendly mechanic. And why should you need one? What do geographers do?

Perhaps you'd need a good geographer on one of those days when everything is off kilter, when the solar eclipse is not on time, or when the wind is blowing-in from the north-east instead of from the south-west. It's hard to know what to do! On days like these you might need your geographer.

But do you know any geographers? Can you describe what they look like? What kind of places do they frequent? They're not the kind of people you'd find in the yellow pages.

I know one or two. You probably know some yourself. They are difficult to spot, like the spotted owl, but at the same time they are in our midst. Geographers are not gender specific nor are they more apt to be found in this part of the country or that. They are not remarkable by the clothes they wear or by their temperament, and yet they are very special people. You may even be one yourself. But how would you know? Does one have to attend school to be a good geographer? Are you a good geographer if you do? What is a good geographer... what is a bad geographer... what is a geographer? There are so many questions about geography and geographers that it is worth taking an in-depth look at them and their profession. What do they do? What can they do?

When people ask me what I do, and I tell them: "I am a geographer," I invariably receive the following observations: "Oh yeah, geography! I remember when I was going to school, we had to learn geography... I loved geography. In fact, I was very good at it. I could recite by heart the capital cities of all the countries in Europe."

What a hero!

What a tragedy! That is a pretty dismal response and a typical appraisal of what the average person remembers about geography. For the ordinary man in the street geography is a thing of the past, a memory from school just like small desks, ink wells and old maps. For him, geography isn't a set of skills, nor is it something that you could use every day, and it certainly isn't a way of thinking for living. But that was then. There was little use in my trying to convince anyone that I had something special, and that they also could have it, and at no great cost. How could I inspire them with the notion that this something special would change their outlooks and their future? What did I know?

Time changes everything, and nowadays geography is viewed in a new light.

Today also, a real adjustment has taken place in the perceived attitudes about many things that pertain to effective education, especially as it relates to geographic awareness. We look with new vision on matters relating to global realities and there is a profound recognition that an informed literate populace is necessary to proliferate and manage this planet into the future. You must agree that this is a commendable idea.

There are serious questions surfacing among thinking Americans about issues like isolationism, world-wide supremacy, global stability and international co-operation. No longer is it a reality that I, as an educated American, can stand idly by, detached from the rest of the world, and think to myself:

"I'm OK, Jack!"

Bad mistake, not anymore! The recent spate of international wars that drew us in and deeply committed our life's blood, should remind us that the global village is at hand and we cannot afford to ignore the writing on the wall.

Did you happen to notice those colorful maps on the same walls?

We've got to face the facts: geography is here to stay! We need it, and our children especially need it, so that they can survive and be responsible contributors in tomorrow's world.

So, are you a geographer? Do you know? How can you tell, and by what criteria can you measure yourself to find out? What are the basic ingredients of geographers? Here are some clues.

By the time you have read and absorbed this book, you will be a long ways to becoming a geographer, and a good one at that. Our hero, who admitted that he liked geography in grade school and who can still remember all the capital cities of Europe is probably a geographer, though not because he has such a fine memory, especially since Europe has changed so much of late. He is a geographer because he received some elementary training and he probably learned some more along the way by association. He may even have become a great geographer.

Consider the following for a brief instant. We have mentioned the word geography in the last page or so, more times perhaps, than has anybody in any book in the last ten decades. That in itself is a revolution beyond imagination... people talking about geography and reading about it.

Congratulations! You are part of this modern revolution.

As a result we are going to make our planet a better place where each of us can live. There are other ways that you can judge yourself to be a geographer. I'm sure that as you read and study more it will become perfectly clear to you what your particular strengths are in relation to the earth.

I love to spend my leisure time in the pursuit of hobbies that utilize the inherent qualities of geography and that promote the preponderance of exquisite places on the face of this planet. It occurred to me that I'd like to share some of these gems with like-minded readers, especially if it helped promote a new study and appreciation of geography among school teachers, and parents. Truly, theirs is the greatest task today. Our teachers and parents must prepare the next generation of explorers, scientists, leaders and individuals, whose assignment it will be to take care of our planet and preserve the innate beauty and quality that was passed on to us by our forefathers.

Somewhere in the body of this book, I have hidden a secret ingredient that you can use to self-test and discover if you are a geographer or not. It is a simple game, and you may wish to play along. Here goes! There is one question that I deliberately left unanswered to stimulate your curiosity.

Of course the first thing that you must do is locate this unanswered question. It is entrenched in the body of the text somewhere. Next, you must come up with the correct answer, before you can safely say to yourself that "Yes, I am a geographer." But will you know for sure? Let me tell you, that in the process of locating the question and searching for the solution, there will come a point when you will know in your heart and soul that you have been smitten by the bug, and you are a geographer.

I want to know that you verified the correct answer, so send it along to me and I will take it from there. I would love to invite you on a short sailing trip in Puget Sound, here in Washington. That would be my way of saying thank you for supporting this geography endeavor.

Hey, it's my book... I can say things like that!

I hope that you can be there for the sail to make it an unforgettable event. You haven't lived until you've experienced the salt-sea breeze in your hair, the Straits of Georgia to the fore, and the panoramic vista of Mount Rainier and the Olympics aft and to your rear.

You will surely meet other interesting people along on this trip too. So read carefully. You could easily pass over this question and you may not readily find the answer, but it is worth the effort to look for it. Keep that wonderful sail in mind, and enjoy the geographic quest. Good luck and happy reading!

So what is a geographer? I am still skirting around the issues, but we are getting closer to perceiving the entire picture. Geographers are everyday people with a particular vision of life.

You don't have to don an Eskimo seal-skin, vanish from your office for a year, losing yourself in the rugged emptiness of Antarctica, to qualify as a geographer, and yet exploration and travel are part of it.

You don't have to search for erratic boulders and exotic flora in sub-tropical climes or equatorial jungles, but geology and botany are also part of it.

You don't have to dive to the bottom of the Mariana Trench or live in the abyss, to witness at first hand, the phenomena of ocean-floor-spreading and subduction zones, but oceanography, plate tectonics and vulcanicity are also part of it.

You do not need a degree in agronomy or socio-economic theory, but life-styles, social systems, economic and cultural activities are part of it too.

And there is more... lots more. There are more ...ology's in geography than in many other disciplines that garnish a lot greater respect, and still the geographer is not really a specialist. The astronomer is a specialist, so is the physicist, and yet there is a great deal of astrophysics inherent in the study of geography. The principle that you need to understand here is that, in order to comprehend some of the complex laws that govern the disposition of our planet earth, we have to contend with parts of many disciplines. It may seem endless, and overwhelming, but believe me, it's not.

The object of study in geography is the earth, but more precisely, the earth as the home of humans. The geographer has to know about people, as they interact with their planet. It's a study of humans in relation to their environment. But that covers just about everything that we do, since it all takes place on the earth. And more, because the earth is within the Solar System, and that Solar System within the universe.

So is a geographer the universal person? Does this mean that the geographer has a little knowledge about a great quantity of subjects? Isn't that intrinsically precarious, since "a little knowledge is a dangerous thing?"

Yes!
Yes!
Yes!

Yes, the geographer is a universal kind of person, who has to look at the global situation and understand a great many disciplines to comprehend our planet. And yes, the person who thinks he knows it all, because he has ready access to a little knowledge, is a shallow person. Remember that education is a lifelong process. It doesn't happen that one day you're 'Ms. or Mr. Nobody' and the next day, you're this famous 'geographer.' Of course, it doesn't happen that way.

And yes, the geographer will have to be a learned astronomer, geologist, industrialist, economist, and knowledgeable in many other disciplines to fully comprehend our universe.

It's not a big deal, however. Geography is an umbrella term for a subject that has to do with human as they relate to their environment. This refers to man - the collective animal - as he interacts with the everyday things that occur on this planet, and with stuff that happens outside it also. The events that have already occurred on, and outside, this planet in the early years of the 90's are ample evidence that there is a critical need for geographers. We are all part of these ongoing events whether we like it or not. It is better for us to know what is going on and what to do about it, than to drift along in blissful naivete through space and time.

We should wear our geography hats when we need to understand how some planetary but everyday things effect our lives and the lives of our children - things like space exploration, global warming, ozone layer depletion, inner city decay, gridlock, fresh air, pollution, fossil fuels, acid rain, the green revolution, famine and a host of other related matters. These issues will not go away. Man's utilization and exploitation of the planet earth has had detrimental impact on almost every aspect of it, with compounding, far-reaching modifications to each future generation.

Geography is the ideal discipline for monitoring the planetary interaction of man. You can keep your finger firmly on the pulse of existence while maintaining an informed outlook on progress and change. Beyond that there are many related fields of geography in which people can specialize and pledge their energies. Many scientists devote their entire being to a particular project. They spend their time developing theories and solutions to age-old questions. And these scientists, too, are geographers.

What an incredible subject for your children to excel at: one that draws on all their talents and skills so that they can come to terms with the reality of our planetary journey. Perhaps geography has been overlooked as the perfect school subject, since you can tie all other disciplines into it at some point. The geography student needs to be able to read, to write, to observe, manipulate numbers, draw conclusions and make accurate, informed presentations about the various findings that relate to our existence on this planet.

Have educators made a grave mistake in the past? Geography, this all encompassing subject, has been lumbered with the social sciences for so long, where it lay dormant and ineffective amidst the other disciplines.

In my book this is the most ideal school subject for our children because they get to experience such a wide variety of scientific principles and because they are forced to express their observations, feelings, opinions and thoughts in relation to processes and experiments that have a direct bearing on our everyday lives. They can then decide, at a later date, which branch of science is more suitable for them, or to avoid - based on personal experience.

Wait until they are ready - since the readiness is all!

Geography provides more than just a sprinkling of facts and lists of cities and features. It is a synthetic discipline. It draws all the relevant pieces together in a unified whole. From this we get a wonderful overview of the life process and the universal truths by which we live.

Education and schooling may come to an end for some people as soon as they get into the work routine, and many people feel saddened that such an enlightening phase of their lives has come to an end. Conversely, some people are delighted that formal schooling is over. However, it is a grave mistake to think that all learning is at an end, because life is the supreme teacher.

Learning is a life-long process!

This applies especially to geography. The tidbit of knowledge that you pick up today will augment and build on the information that you already know and provide the basis for more growing tomorrow. The good news is that learning will continue to grow and flourish until you give up. The choice is your's to make. You must decide that you've had enough and you are the only one who can pull the shutters down to prevent any more knowledge getting in.

It's all up to you. You're in the driver's seat!

Throughout your life you can proceed to any level of learning and discovery that suits your soul. Geography is a wonderful way to let your mind grow and accumulate facts and knowledge about our world and about man's exploitation and survival in our world. Geography is a way of looking at and experiencing life. At times I wonder how people make it through life without that special vision and perception that a little geography provides.

The morsel of geography that you can glean from this book will be but a grain of sand in the ocean of life, but let it whet your appetite for a greater immersion in a learning that will color your outlook, so that the planet and the world around you, will brighten up and become alive.

Mine is not the only voice that cries out for recognition for this versatile discipline. Today, more and more educators and parents are coming to realize the intrinsic value of the skills and attitudes that are imbued in people by an in-depth study of geography. National and international bodies that wield a great deal of political and cultural clout are finally awakening to the new world of man and his environment. The results are going to be spectacular and will enhance our lives and your children's education for a long time to come.

Mark my words well; geography is here to stay!

A most welcome injection of energy and capital has recently been proffered under the auspicious direction of the National Geographic Society. There is no doubt that the guiding forces in this organization are committed to the long term preservation of geographical studies for the future. In the past few years, we have heard many great speeches and seen equally great gestures that indicate their preferred direction and approach. And I am confident that with their supervision and help, the future of geographical education in this country is in good hands.

This is a reassuring thought for me, but I do not want to give the impression that the job is done or that the battle will be easy. On the contrary, it is my desire and this book's focus to illuminate the method and the direction that we must all utilize in order to bring geography into the position that it rightfully deserves in the nation's homes and schools. In this way, will we be able to justifiably say that we are providing the education that our children deserve. Listen to the words of the president of the National Geographic Society. Stand up, pretend you are Gilbert Grosvenor, read them aloud, and hear their true meaning.

> "We are a nation of people with worldwide aspirations and involvements, a nation whose global influence and responsibilities demand an understanding of the lands and cultures of the world. But, geography, once an integral part of a child's education, has lost its foothold in American schools."

These words mark a new phase in the development of geography education in our country. They denote an era of invigorating assistance from an international organization as it prepares to enter the educational fray and contribute in a meaningful manner to the schooling of our nation's youth. But read on. Consider this next statement that sums up the whole mess and points to an increasing problem in our schools.

> "It's a good thing our ancestors found their way to America, because today, twenty-four million Americans, one in seven, can't even find the United States on a map of the world... one in four eighteen-year-olds couldn't find the Pacific Ocean."

What an incredible dilemma! How could a situation like this come about? Just think about it for a moment! One American in seven can't find the United States on a map of the world! One in seven! That means that if you are in a room with six other people, one of you cannot find the U.S. on a world map. Look around you. If the other six appear to know what they're doing - then you're it! What a comforting thought.

It is easy for me to accept that our children must receive a better geography education. This is because I am sold on the idea of a clean, healthy environment and a safe planet for the future. But I also think that our future leaders need to be able to understand our environment and be aware of global issues in order to better compete in the world marketplace. It is a principle of critical economic importance for the well-being of our nation in the years ahead.

And yet we read almost daily about students who graduate with little or no knowledge of geography. George Bush, who wanted to be known as the *education president*, focused at length on the problems associated with a generation that turned out illiterate in geographical skills. He called for a reappraisal of matters geographical and educational.

"When some of our students actually have trouble locating America on a map of the world, it is time for us to map a new approach to education."

I see a deeper malady hidden within this disclosure. The average person reading this negative piece of news will continue to go about his daily chore, dismissing the article as referring to one of those long haired, barbaric looking youths who cannot locate the Persian Gulf, or Nicaragua, on a map of the world. And that is that! Well, what did you expect - look at the way they dress!

This is a sad indictment on our educational system and particularly devastating for our teachers, who are accepting the brunt of the abuse on the front lines. But these accusations can be made and substantiated. It is true that many of our high school graduates have neither the skills nor the ability to locate places, countries, climatic regions and other relevant data on a map of the world. Moreover, many also fail to comprehend the various physical intricacies that cause our planet earth to behave as it does. Do they know why the seasons occur? Why is London eight hours ahead of Seattle? What causes the midnight sun?

But this is only part of what's wrong with geography in American education today. I cannot blame the students - they will learn what they are given to learn. And I cannot blame the teachers. It is wrong to lay the blame for the nation's geographic woes at their industrious feet. They already have enough on their plates. And in spite of all the negative publicity, I believe that teachers do an incredible job dealing with young people in today's media crazed times.

Teachers have to compete with the graphic immediacy of Nintendo, the gratification of global screen images at the touch of a button, the 'we must have it now' syndrome. And all this takes place against the social background of today's functionally maintained family structures, such as moms working, single parenting, drugs and alcohol, teenage suicides, Aids and the threat of global destruction because of nuclear capacity. Their's is not an easy lot. No! The onus for the poor plight of geographic education in America today, does not reside with the teaching community.

We have to take a deeper look at the historical circumstances on a broader basis, to understand why geography education is the way it is in our schools at this time. In so doing, and by comparing what has happened here with what took place in some other countries it is possible for us to get a pretty accurate overview, and a comprehension of the situation that has occurred. And more importantly, this comparative view can offer us a template for future improvements. After all, we don't want to reinvent the wheel. We can learn from other nations and utilize eclectic decisions in our selections into the future. We can choose to include the things that seem practical, omit the issues that don't work, and avoid anything that was instrumental in causing strife.

In the next chapter, we will look back a hundred years or so, to see how other countries dealt with their educational systems and learn what they did to arrive at the pinnacle of success today. We will particularly focus on countries that are in the western culture and have proven that geographic awareness is at a greater standard than in our nation. From this we will learn how best to foster a new growth in geography education in our homes and schools across the land.

Chapter Two:

The European Lesson

The real kernel of what ails geography in American education today is etched in the mores, attitudes and opinions of a century of mis-informed people. The ordinary man-about-town is no different from the hero mentioned in our last chapter, who thinks that a geographer is somebody who can point to Saudi Arabia on a world map or who can rattle-off the capital of Turkey from memory. This is the real issue.

Therein lies the answer to the question - "What's wrong with geography in American education today?" The average person, who may think he has the solution, hasn't got the faintest idea what geography is about. Not a clue! How could he? His teachers before him didn't know. Map pointing and memorization per se do not comprise geography! I cannot stress this point enough. The ability to point to places on a map, or to spit out the names of capital cities around the world, is not to be derided. It means you have a good memory and maybe an aptitude for travel. But it is not geography.

Geography is not a list of capes and bays!

In order to fully comprehend this critical premise, we need to don our comparative hats and take a quick look at the educational systems in some other parts of the world and see how geography developed there. The first thing to realize is that geography is alive and well in other countries, and especially in Europe, where there are very good historical reasons for its status. This did not come about by accident, nor did it happen overnight. On the contrary, there has been a forceful drive to bring geography to the forefront of educational philosophies and keep it there, since the middle of the last century.

Geography, as a scientific discipline, underwent a major revolution in Europe during the last decades of the nineteenth century. Geography, as an educational process, became accepted throughout the burgeoning school systems in each country, not as a segment of another discipline, but as a scientific study, unique in its own right.

As education spread and was accepted in America, it is unfortunate that the new geography was not incorporated from the beginning. That is why, until recently, geography in America still resembled attitudes that were prevalent in Europe about one hundred years ago. We have to admit it.

Geography is more advanced in Europe today than it is in the United States!

But not for long! That is up to you and me! To understand the development of geography education in America it is necessary to view the entire historical tapestry. At a time when geography became firmly established as an academic discipline in the schools and colleges of Europe, many factors, emanating from the diverse nature of the American continent, coalesced to militate against a similar growth to fruition on this side of the Atlantic. This is where the educational endeavors of European schools differ from American schools. As education developed in Europe, geography grew along with it, but here in the U.S. education was linked to social issues so that geography was left behind.

Other factors, too, helped the immediate adoption of academic geography in Europe, factors that were not so apparent in America. After all, Europeans were approaching the end of the 'Ancien Regime' an antiquated system, decaying and crumbling, whereas Americans were at the beginning of a new life-style, tentatively exploring their way around democratic principles of self-government and education. There were a thousand new things to attend to every day.

The results were predictable. Geography was adopted as an important academic science in Europe, but it was largely overlooked on this side of the Atlantic. Why was it that Europeans accepted geography and Americans did not? Is there something that we can say was the root cause for this turn of events, resulting in the present situation? A closer inspection of the political and social climates in the two continents reveals the underlying factors that explain this occurrence.

When we study the educational events that took place in Europe over the past century, many things immediately become apparent to us. We discover uncanny analogies that may be drawn between the plight of geography education in Europe then, and that in America, now. It is mind boggling to determine that many of the goals and objectives that we are attempting to accomplish in America today, were the same missions fought and won, over one hundred years ago in the school rooms of England, France and Germany.

The work of enlightened European individuals, many years ago, can be equated with the tasks being presently undertaken by the National Geographic Society and other educational bodies, which are working for the betterment of geography education, in our schools. It becomes obvious to us, that geography education in America today is at a similar position to what it was in Europe, more than fifty years ago. But what was going on in educational matters at that time, that makes it so influential today?

In Europe, the educational principles of Jean Jacques Rousseau and Pestalozzi were incorporated in the professionalism of such geographers as Karl Ritter, Elysee Reclus, Andrew Herbertson, and James Fairgrieve. Because of this simple ideal, all teachers received ample and adequate training in the rudiments of geography teaching method. The work of a few men contributed to the overall adoption of geography in the classrooms of all the schools in Europe.

Wait! Who were these people? What does all this mean? You may well ask - who were Rousseau, Ritter, and the others and what was their connection with geography and education?

Jean Jacques Rousseau is well known in educational circles on both sides of the Atlantic. He was probably the man who most influenced education, in democratic states, in Europe and here in America. A theorist and philosopher, he laid the precepts for the French and American revolutions. His ideas on liberty, equality and freedom form the basis for much of modern man's thinking. Two books, that are attributed to him, particularly affect education in a democratic nation.

"Man is born free, but everywhere he is in chains"

He screamed these words from the pages of *The Social Contract* as he strove to liberate man from the state of the savage and introduce him to a civilized existence. In his other book *Emile*, he forged the foundations for the education of the individual - an education that was both human and humane, and fitted to the enlightened man of the new world.

Pestalozzi was the greatest teacher, though less well-known in America. He differed from others because of his hands-on practical approach. In eighteenth century, war-torn Europe, he gathered the needy, homeless orphans and showered them with caring, love and protection. While engaged in this lifelong pursuit, he uncovered the elements of erudition, cognition and motivation. He experimented with the **'Look and See'** approach to schooling and it worked wonderfully. Pestalozzi's importance lies in the fact that he instigated the first teacher training college in his Ecole Normale. He was the fountain-head of teacher training, the vital link between the academic and the layman.

Karl Ritter was a German geographer who visited Pestalozzi and was so influenced by what he saw and learned that he devoted his life to putting Pestalozzi's method to work for geography. He assembled the hitherto compendium collection of scattered facts and systematized them into a new and practical science. Ritter was the founder of modern geography.

Elysee Reclus was a French philosopher who realized that geography was an essential discipline in human understanding. He worked ardently to bring geography to the loftiest academic heights at the university.

Andrew Herbertson and James Fairgrieve were practitioners who brought geography from the ivory towers of academe to the teacher, and beyond to the layman. Fairgrieve is famous for his virulent adage that summarized the principles of a common method of teaching before outdoor Field Centers were set up in Britain.

"Geography should be learned through the soles of the feet."

These were great thinkers, philosophers, and geographers who realized that education was the basis for successful people and successful countries in a democratic union. Through the energy of such people, geography has achieved a lasting and comprehensive position in the lives of everyday people in Europe today.

From the outset, it was recognized that knowing **how** to teach was as important as knowing **what** to teach. Method was the key to transmitting the contents of the curriculum. This differs from the commonly held belief, practiced in many schools today, that if you, as a teacher, know some general facts, can interpret a map, explain how rain occurs and so on, you are a prime candidate for teaching geography. Not so!

Wrong!
Wrong!
Wrong!

This is a serious mistake. Too often we have seen subject experts, scientists and intellectuals, who are quite brilliant at their work, but who make terrible teachers. Unfortunately, competence at a particular subject area does not automatically translate into great teaching skills. People either are teachers or they're not. There are no two-ways about it. Some people cannot teach and never should attempt to do so. It is a travesty of the greatest magnitude to see young people pummeled and lambasted by people who are supposed to be teaching them.

Geography teaching methodology is an extremely specialized science and a study in itself. It is a very necessary component in the imparting of the scientific method that is geography. This was one of the primary reforms that came about in Europe as a result of the growth of national systems of education and the development of education as a process. The same principles are just as important in America today. It is essential that all our teachers must receive the basic rudimentary elements of geography teaching before it can be successfully transmitted to the students with fervor and enthusiasm.

It is not enough to know geography - it is also essential to know how to teach it.

Why did educational method proceed in this enlightened fashion in Europe and why not here? The answer is pretty obvious if you allow your imagination a visit back to a country like France in those times. The order came from above.

Europeans were accustomed to accepting orders from the top. Though democracy was an ideal for which they fought, they still had their Kings and Queens. Governments continued to lead the people. The old regime persisted, though altered considerably to reflect the French Revolution, the Industrial Revolution, and the growth of Labor Unions. Customs and mores that had been culturally imbued on European psyches for centuries could not be discarded overnight. They were still part of the rational fabric of the average European. Therefore, once geography was accepted by the universities at the top, it easily penetrated to the grass roots.

Teachers were trained at the universities and became proficient in teaching method as well as geographical skills. They first attended the academic school, where they were given the knowledge and shown the processes in the science that is geography. Then they attended the school of education where they were honed in the method, and supplied with the tools and techniques for imparting geographical skills and knowledge to their students.

Geography did not always enjoy such a prestigious position in the schools of France and Britain. But as the old order began to tumble down, and national systems of education developed, there were sufficient informed and dedicated individuals who focused on the issues and fought obstinately for geography. We owe much to their vision and drive because today geography is in its rightful place in the curriculum of the new schools.

Let's visit back at the beginning for a brief interlude, and learn where it all started. This may seem to you to consist of dry historical data, but believe me, it's not, and a quick perusal of the origins of geography education in Europe will give you a breathtaking view of your own position today. So hang in there!

Geography education began as an exploratory hobby in Britain at a time when colonial expansion was placing adventurers in many exotic regions, and unforgettable new discoveries were unfolding daily. In 1830, the Royal Geographical Society was founded in England by members of the Raleigh Dining Club, whose previous and rather dubious claim to fame was, that they had collectively visited nearly every part of the known world.

The Society was a direct follow-up to the African Association and was equally dedicated to the promotion of exploration on that continent as well as elsewhere.

The Royal Geographical Society's initial years coincided with such exploration luminaries as Livingstone, Burton and Speke as they were laying bare the age-old secrets of hitherto unpenetrated continents. Burton and Speke's exploration has always been popular and today it is well documented in books and even a movie called *"Mountains of the Moon."* What a wonderful time to have been an explorer!

At that time also, the economic dynamic involved in imperial expansion was affecting geography at the highest level. The Royal Geographical Society members played a key role in this sphere since they directly linked the significance of imperial expansion overseas to economic well-being at home. Burton's journey to Mecca and his attempts to find the source of the Nile were intrinsically linked with enhancing the economic value of these imperial territories with the growth of British mercantilism. In Africa, for example, with the British Empire established in the north of the continent, in Egypt, and in the south of the continent, in South Africa and Rhodesia, the economic vision was to link the two ends of the continent with a railway system and eventually a water system.

Initially, education was not a concern of the Royal Geographical Society - the watchword was 'Exploration not Education'. It was inevitable, however, that the question of geography in education would arise as more people were required to maintain the colonial expansion. As a result, Captain Maconochie, of the Royal Geographical Society, became the first Professor of Geography in the UK, when in 1833, he accepted the post on the invitation of the University of London authorities. His appointment was a milestone of great significance to geography education.

In the first place, since the University of London was a degree-granting body and geography was recognized as an academic science, there was now an incentive to work towards a qualification. Secondly, Maconochie's appointment maintained the close link with exploration and discovery by sea, and perpetuated the practical, inductive method in the field.

As academic sciences were becoming entrenched in the universities, there came about a gradual change in emphasis from the world of exploration to that of education. Geography education was flourishing on the continent, both at the universities and in the schools. Under the influence of Karl Ritter and his pupils, geography was well established in Germany, while in neighboring France, Elysee Reclus, a disciple of Ritter, together with Vidal de la Blanche, put geographical education on an equal footing with the other disciplines and sciences. Globally, a greater volume of geographical knowledge, resulting from the opening up of both the new world and the interior of Africa, was available than ever before. The time was ripe for a greater involvement of geography in education, especially in Britain, where so much of the exploration and discovery originated.

But everything was not right. The plan was not operating as smoothly as people at the Society wished. Then in 1884, an able Scotsman and active geographer, John Scott Keltie, was commissioned to undertake a fact-finding mission into the state of geography education in Britain and the continent. This was the second most important milestone in the development of geography education in Europe.

Keltie was instructed to visit Germany, Austria, France, Switzerland, Italy, Belgium, Holland and Sweden and to make inquiries by correspondence as to the state of geography in America. He was authorized to collect from every country characteristic textbooks, maps and appliances of every kind for the teaching of geography. He spent a number of years in the pursuit of this quest.

Keltie's study was commissioned by the Royal Geographic Society, and was published in the year 1888, in England. Look at this! At the same time, in the same year, the National Geographic Society was founded in this country. And it had the same stated resolve, that of increasing and diffusing geographic knowledge. Hardly a coincidence! This is where education was at in those days.

It is obvious that geographical awareness was a prime focus in the last decades of the nineteenth century. The historical development of the two geographical societies is ironically similar in character. Here also, education was not a major objective of the Geographical Society at the beginning. They may have started in the same position, but before long they had drifted-off in very different directions and away from one another. We will follow their separate paths for a little while to glean some comparative facts so that we can better plot a direction for the future of our educational endeavors.

Keltie's study revealed the incredibly poor state of geography education in Great Britain. In contrast with the other European countries, Britain was by far the worst in terms of equipment, commitment and method.

Doesn't that statement ring a haunting bell? Today, American students are being contrasted with Japanese, Asians, Australians and Europeans to demonstrate that America is by far the poorest in terms of geographical awareness. But that was in the 1890's, and this is the 1990's. A hundred years has passed by! What happened in the meantime? It's worth going back to see what transpired after Keltie's commission.

The council members of the Royal Geographical Society were embarrassed, and reforms were the immediate objectives. They embarked on a prompt campaign against the existing institutions to instigate some changes. A very simple and ingenious plan was devised.

It was understood that if Oxford and Cambridge established Chairs of Geography, teachers could be adequately trained and reforms would spread down to the schools. This pyramid principle was the plan. Geography could be introduced at the top so that it could percolate down to the second and elementary levels of schooling. Teachers were the key agents in spreading this new learning.

In England teachers were the key.

The result was that Chairs of Geography were set up and standards of intellectual achievement rose, soon after. Naturally there was not an immediate abundance of trained teachers and many different methods were adopted to increase the numbers of available personnel to manage the task. The most successful was the utilization of the summer vacation time to do intensive training in the content and method of geography teaching.

Field-work centers opened and flourished as it was perceived that experiential education was a welcome addition to turning the learning process into a motivational and meaningful event. Year after academic year, more and more students were graduating with skills in geography, and a love of their subject. Some went back to teaching and others went about other daily tasks. But all were armed with a geographic knowledge that prepared them for living on our planet.

As the educational system in Britain grew and flexed in response to historical change, the curriculum always embodied geography because it was part of the basic premise, as encompassed by the educational philosophies. The drive for equality of educational opportunity and quality education in the wake of the eleven-plus selection system during the years following the second world war meant that more and more people were being exposed to geography. When Comprehensive Schooling was established in the 1970's, geography was a full subject in the school curriculum. Today every student receives adequate training in the craft and process of geography; training that is competent and effective, carried out by skilled teachers.

The publication of Keltie's report marked a new era in the development of geography education in Great Britain. From that day forward geography was working its way into the school system and insuring its place on the curriculum. Now, let us compare the development of geography education in Britain with what happened in this country. The next chapter will briefly look at the political and social factors that resulted in the situation we are faced with today.

Chapter Three:

The American Experience

In the last chapter we saw how geography and education developed and progressed together, during the last century, in countries like Great Britain, France and Germany. This explains why their students and the ordinary man in the street each have a solid working knowledge of geography and more than a passing acquaintance with the intricacies as to why our planet behaves as it does. We can learn a great deal from an inspection of the process and the activities that occurred as the system of education was growing and as geography was becoming adopted as a reputable and vital subject on each school's curriculum. We are at a time right now where geography could either "go", or not, in our school system, and I would like to see it become successful. We could all benefit from a positive application of geography to our lives as you'll discover while reading this book, I hope.

We saw that there was a critical turning point in the development of geography in education in Great Britain about one hundred years ago when the Scotsman, Keltie, released his findings as to the plight of geography in the schools across the continent. I think we are at that point now in our own evolution, and we should start from there and make quick strides to catch up with the rest of the world. The first thing to do is look at where we are at, and then we must rectify that position.

We don't have to go back very far to get a direct link with Keltie's report. In fact Gilbert Grosvenor made a report that was just as dismal only a few years ago. This was at the celebration of the centenary of the National Geographic Society, in the year 1988. Yes, nineteen eighty-eight, one hundred years after its inauguration, when they brandished those lofty aims, "for the increase and diffusion of geography." In his speech Grosvenor proclaimed the Society's failure to bring geography to the ordinary man. As bad as the situation may have been this was a speech of utmost significance. With this disclosure, geography education in the United States had reached a critical water-divide.

I would say that this was probably the most momentous event since the Society's founding a century before that. It has more significance than epic voyages to the poles or daring ventures in outer Mongolia because it begins to bring the world of the geographer into the livingrooms and the classrooms across our great land. And it makes an attempt to re-address the real issue of geography, which is, man in relation to his environment.

There were many reasons why geography education did not flourish and abound here in the same way it did in Europe. A serious look at the growth of American education, the rise of democracy, and the spread of new ideals for a progressive society will help us understand how geography was overlooked and why it ends up today a poor cousin, lying limp, at the bottom of the educational mound.

At the outset it seemed that education and geography would go hand-in-hand in America. Instead, it appears that education and democracy were soul-brothers. This, of course, was not a bad thing.

All political ideals require educational vehicles to transport them to their goals.

Schools are important to governments; and this is demonstrated time and time again as educational growth comes about in response to historical change.

In times of need governments will call on the school system to bring a particular skill-set into focus. This occurred in 1957 in the wake of the Soviet launching of Sputnik. Government intervention created a dramatic effect on the development of science and math in this country. The same happens to subjects like physics and computer science in times of war when the demand for technicians and operators is great. Today's technological age is a prime example of how the work place dictates the educational requirements, and influences our schools.

Educational enhancements have, at all times, been responses to historical change.

Similar trends occur today where the need for skills in technology is apparent in our hi-tech society and this need drives our educational philosophies. At other times the education process is used, with good effect, to maintain the status quo.

Rousseau's and Pestalozzi's ideas did successfully cross the Atlantic and became part of the schooling system in many areas. Ironically, the democratization process advocated by these men was one of the reasons that geography never had a real chance to be universally accepted. In a democracy, universal acceptance can only happen by the majority consent. That consent does not happen haphazardly and needs focused and concerted attention to bring about the desired situation. And it so happened that there were always too few people sufficiently inspired with the zeal of education and geography to successfully infuse and manipulate the growing population of the new and expanding country.

But it is to the philosophic principles of another European, John Locke, whose ideas also made it across the Atlantic, that we must look to glean an understanding of the development of education and democracy in the fledgling nation. Although abhorring the excesses and prejudices of the ancien regime, the new world philosophies and ideals did not spring from the ground overnight to suit the great land. They were, in fact, borrowed from the trials and tribulations of the French and English systems, from men like Hobbes, Locke and Rousseau. These philosophers were colossals that spanned the boundaries of both worlds, and their effects would be felt equally in the French and American revolutions.

Influenced by the universe of perfect motion that was unfolding under the vision of men like Galileo, they now introduced man, the former savage, to the microscope of scientific scrutiny. Galileo had already liberated mankind from the shadows of mythology and superstition, and now it was time to help him into a world of civic order. Locke and Rousseau, like their predecessors Plato and Aristotle, theorized on the age-old question concerning the nature of man in society.

In the Europe of the feudal system, serfs did not own property and their rights were governed by their courtly lords. It was a time when people were destined to be ruled by church, by king and by noble. Wars and misery were widespread since nature was a poor provider for mankind. As the population increased, it was easy to see that the basic necessities for life and survival, including food, clothing, shelter, and health, were not commodities that abounded in plentiful supply. Man was forever protecting his meagre possessions from his fellow-man. The result was war, strife and pain, each culminating in the very real threat to life itself. Self interest was anathema to civilized man, but enlightened self interest could work. People united in a system of logical truths could assist each other to attain new heights in their quest for security and prosperity.

Rousseau, therefore, proposed a social contract, whereby men would come together sharing a common purpose in a civil society, to rectify the human problem and eliminate scarcity and universal war. Nature was isolated as the enemy of prosperous man. Scientific discoveries and evolving technology could liberate him from the slavery of seasonal dependence and climatic disasters and, thus, became the friend of modernity. He would use the powerful knowledge of science to wage war, not on fellow-man, but against his common enemy, nature. In this way, people, who were vulnerable from the exigencies of the world, could unite with their fellows and become the conqueror. Together, then, they could provide for themselves and achieve a state of prosperity and happiness by ending pain and strife.

Locke took this philosophy a step further. His utopian existence would impart these privileges to all men - privileges which had hitherto been the preserve of the birth-righted few. For the first time in history all men were to be treated equally, and for the first time in history all men could own property. These were revolutionary words in a time when revolutions were needed. Thomas Jefferson was no stranger to Locke's book, *Two Treatises of Government*, and it is no accident that the same rights Locke accorded to all men as inalienable, appeared in the Declaration of Independence. Locke had said that by nature all men were free and equal and that they had the right to life, the right to liberty, and the right to own property. These sentiments were exactly what Jefferson needed to hear when it came time to draft the declaration that would end the meddlesome interference of England in the colony's local affairs.

In order to live this life with peace and prosperity, man needed to be competent at the application of science so that he could conquer nature. Education, and especially a national school system, was seen as a primary key in the creation of this kind of society, because without the diffusion of knowledge among the people, the experiment at self-government was doomed to failure. Jefferson preached a crusade against ignorance and championed education as the only sure foundation for the preservation of freedom and happiness.

The establishment of the system of democratic principles, the drive to produce and contribute to the relief of human misery, and the goal to become nature's conqueror, were more than sufficient to fill the plates of even the most ambitious rulers and educators. Where was there room for geography? It couldn't possibly compete when such loftier goals were viewed next to it.

The American pragmatic attitude—"if it works, it's right"—helped explore, tame and inhabit a hostile continent. Self-governing people were needed to colonize, inhabit and populate the vast, opening continent, as they inched their way from the initial toe-hold along the eastern seaboard, across the rolling prairies, over desert and mountain to the far-reaches of the Pacific coast.

In the early days of the nation, the inalienable rights of mankind, the right to own property, personal autonomy, freedom of the press, religious liberty, and citizen obligations were crucial factors for the effective functioning of a republican government. Life in the new world was not going to resemble that of the world that they had just abandoned. The political tenets were founded on the principles of freedom and equality by the consent of the governed. And this was all to be made possible by advances in science and technology that would conquer and tame the natural wilderness to provide prosperity, health and happiness.

During the first half of the nineteenth century, American statesmen turned to education as the panacea for the country's social, religious and economic problems. Abraham Lincoln averred that education was the most important subject for Americans to be engaged in. Thomas Jefferson was adamant that freedom and ignorance would be impossible bed-fellows.

Education was the key in the world of life, liberty and the pursuit of happiness.

As towns and trading posts grew up, schools were built and the task of educating the young began. The common schools were entrusted with a formidable and onerous task - that of inculcating the new generation with civic, American and moral responsibilities.

In the late 1800's America became asylum to the millions of oppressed, tired and hungry immigrants that fled European countries. Local problems of ethnic origin accompanied each new wave of migration. The task of Americanizing these immigrants was turned over to the educational system—the common school. It was necessary to transform Europeans into loyal American citizens—else there would be 'little Italy', 'little England', 'little Ireland' and more. The process of assimilation further taxed the education system. But it worked—so it was right!

Lone voices continued to cry out about the plight of geography in the schools. Textbooks and equipment were needed. After Independence there was a perceived need for geography texts with an American slant. There were plenty geography books written in English, but they were British. Anti-British feeling was high, a factor that lead to a bizarre turn of events. An American clergyman, Jedidiah Morse, perceived the demand and wrote a series of text books entitled - *American Geography*. Morse had more business acumen than geography and his publications were enormously successful, reaping a rich harvest of dollars—the first in educational publishing. Morse's texts were terrible, steadfastly national, narrow in matters pertaining to religion and ultra-conservative in moral tone.

His *Geographies* were merely literary accounts of geographical phenomena and were devoid of any intellectual stimulus. They were dramatically successful, nevertheless, and spread throughout America rapidly in a time when Americans were eager to learn about their great land. As a result geography suffered a severe intellectual set-back and was dropped as an academic subject from the curriculum of most colleges in the latter half of the nineteenth century.

This created the peculiar anomaly whereby geography flourished in the halls of academe in Britain, France and Germany, while it virtually disappeared from intellectual habitats in America. What had worked in Britain could not achieve any lasting results on this side of the Atlantic. The simple method of transmitting information from the top down, pyramidal fashion, was thwarted, because democratic man hated to get anything in this manner. Teachers were not trained in the skills necessary for the effective teaching of geography. Attempts lead usually to the same end—rote learning and memorization of long lists of Capes and Bays. Innovative ditties and anagrams were invented to help with the retention of dry and uninteresting facts - but these were no substitute for the real thing.

The lofty words of Socrates, Pestalozzi, Ritter and Rousseau fell on barren ground.

As is always the case, there existed a small number of enlightened educators and philosophers who labored against all odds to bring their message to the unfriendly world. William Morris Davis was one person who did much to foster the advancement of geography in America and Europe. It was his interpretation of Geography—environmental determinism—that dominated American geography for the greater part of the century.

Davis was, like his contemporaries in Europe, an avid outdoor geographer. In 1910, he was in Ireland with Grenville Cole studying the superimposed drainage pattern and the strange right-angled bend, discovered by the renowned geographer Jukes, on the River Blackwater. For him, there was no substitute for raw information, made from direct observations and the thrill of discovery associated with field-work. He was a strong advocate of hands-on, outdoor work for school students.

But the task was too enormous for one enlightened individual. He was an important link in the chain of events that kept geography in the forefront of some university courses since his students, in particular Mark Jefferson, carried the message to many more, keeping the flame alive.

There were others who were infused with the zeal of geography and worked tirelessly to promote it at the universities and schools. Ritter had taught many brilliant geographers who, in turn, carried the torch to new students. Arnold Guyot became professor of Physical Geography and Geology at the College of New Jersey in 1854. Guyot, and two other pioneers of educational methodology, Barnard and Mann, were luminaries in an age of change and turmoil. Their work should have been rewarded with greater success, but the burdens imposed on the education system were too excessive. The time was not right for the diffusion of geographic knowledge in the national school system.

Dedicated geographers did not give up, however, and concerted efforts were maintained to promote the study of their favorite discipline at local and national level. The Association of American Geographers was founded in the year 1904, and worked tirelessly to install academic geography in the Universities across the nation. Similarly, the National Council for Geographic Education was founded in 1915 to promote geography in all our schools.

In many countries geography became coupled with history and civics to form a more manageable curriculum. In America it was known as "Social Studies." This came about for numerous reasons, all having some practical good qualities. But in all cases the effect on geography was detrimental. The umbrella of history and civics diluted a science that encompasses our lives.

That geography should be treated as a viable academic subject in its own right was a question that many people were prepared to tackle and defend. But the years during and after the first world war were hard times. There was an economic recession, followed by another major war. National security and financial woes seemed more important than academic justice. And maybe they were right. We have to prioritize our problems. But the end result was that these and other social issues of considerable import usurped the time and energies of the people in control.

The American Geographical Society took up the cause against social studies by making a call for a return to geography as a scientific discipline. The members made presentations to educators seeking support in an attempt at rectifying the matter and raising geography to its rightful place in the classroom.

Testimony on the pages of the various journals bear witness to the many fine educators and geographers who championed the cause of better geography teaching. Geography, according to these educators should be as inspiring as opening up new frontiers, reaching the poles for the first time, or climbing the unclimbed peak. Foremost among them was Isaiah Bowman, a disciple of Davis and student of Jefferson, who preached a crusade against drudgery and repetition.

Lack of teacher training continued to be a problem. The universities were ill-equipped to adequately train teachers in geography and there was little demand for such training. Through the late forties and early fifties no major improvements occurred for geography in American education. The gap that existed between the expectations demanded of education and the stark reality in the classroom grew greater as new issues raised their menacing heads and old problems lurked ominously in the background.

The civil rights movement highlighted the real truth concerning the egalitarian ideal - equality of educational opportunity. Poor people and social and cultural minorities had genuine misgivings about the nature and quality of their education. The educational horizons of a child were controlled by where he or she lived. Educational advocates focused on these issues and worked to bring about the reforms that help improve the basic schooling that every child deserves regardless of his/her background.

To the student of history, the picture is clear. Sometimes the spotlight was on civil rights, other times on educational opportunity or equality, but never was the focus on geography. For this reason the post-war period saw little improvement in the position of geography education in schools.

National and global issues maintained a formidable presence in the forefront of educational endeavor. Literacy levels needed to be constantly upgraded in the new era of technological advance and burgeoning artificial intelligence. Educational experts were overloaded with poverty programs, equality questions, federal and local issues and other economic and social matters that occluded the needs of geographers.

The year 1957 was an important one for the development of American education. That year witnessed the launching of the Sputnik satellite, an event that relegated the United States to the back-runner position in the space race. This reality had an immediate impact on education.

Federal grants were made available in the National Defense Education Act of 1958. At first, science and mathematics were highlighted as the focus for attention, but later, in 1969, geography was added. However, it was a long, arduous, and uphill battle to bring geography to the forefront of educational thinking, and before the public eye.

But change must come. Two organizations, the National Council for Geographic Education, and the Association of American Geographers, laid the foundations for the sweeping reforms that were about to take place. In 1982, Gary A. Manson of Michigan State, and Richard L. Morrill of Washington State, headed up the committee that compiled the blueprint for the *Guidelines for Geographic Education in the United States*. These guidelines formed the basis of instruction to be used by teachers in the schools across the nation.

These were valuable and important developments for the provision of geography education, but they were not enough. Still, geography was a second rate subject with no status of its own. Teachers were inadequately trained, if at all, and the majority of our students were graduating with little or no geographic skills.

Help sometimes is enlisted from unexpected sources. A new force was coming into play that would eventually help bring the poor plight of geography education before the nation's eyes. This new catalyst was public opinion. Fed by the enthusiasm of the media, it has played important roles in, and has been an intricate component of all political movements in recent American history. The circumstances that lead to an awakening of geographic education standards were entwined with the military activities that occurred in the 1980's and 1990's.

A time of war throws added light on geography since new and exotic names appear on our newspapers and TV screens - not to mention maps, charts and global pictures from space. This first happened after Granada, and grew to a special force during the crisis in the Gulf. Suddenly the press became aware that the general readership was ignorant regarding the location and lifestyle of some region where American soldiers were expending their energies, and often their lives. Then the realization sank in that there was a flagrant lack of knowledge among the public at large concerning our geographic wherewithal. Spurred on by the triumph of their initial success, like Thorndikean cats, the media folks heaped more and more exposure on the base ignorance of the nation's youth. The educational system was, of course, the culprit, and another call to "bring back the basics" ensued.

This type of democratic consent can often be harnessed to bring about reforms if the appropriate ingredients are present. A strong unified body is needed to take control of the surge and channel the public enthusiasm in a coherent and manageable direction. Timing is critical. In 1988 that strong unified body was about to celebrate its centenary, in recognition of the notable progress that was achieved since the founding fathers stated its lofty aims. The National Geographic Society was one hundred years old.

At that event education experts and geography professionals assembled in unison to celebrate their many achievements, but also to decry the deplorable state of geography education in the schools across the nation. Wouldn't Keltie have enjoyed that meeting?

It was a profound gathering, and a veritable coup for geography education in this country. At last there was a clear focus and a stated mandate for the future. The proficiency and expertise was demonstrated to be willing and available. The urgency of the situation was adequately exposed, impressive plans were drafted, and speeches were delivered, both meaningful and sincere. Wouldn't you love to have been there, to hear Gilbert Grosvenor voice these crucial words?

> *"There is no more fitting way to begin our second century, than with a renewed and expanded commitment to Geography Education. The Society's concern about the untenable consequences of geographic illiteracy compel us to take an even larger role in education, and we are in it for the long haul."*

And long haul it would be. One of the keys to students receiving a better geography education was, and continues to be, teacher training. Parents have an equally important role. But the National Geographic Society's mission to improve geographic literacy began and ended with individual teachers. They supported teachers across the land with a range of services designed to enhance their knowledge of geography, and their teaching skills. One of the first things that the Society did was set up the Geography Education Program to provide a permanent and expanding base of financial support for exemplary geography education designs.

They made a permanent commitment to teacher training, considering it a vital element in the drive to improve the status of geography in our schools. They did this in many ways. First, they mobilized a wide range of Society resources to improve geography instruction, focusing these resources in five strategic areas, namely, grass-roots organization, teacher education, materials development, public awareness, and outreach to educational decision-makers. Then they established the Geographic Alliance Network to get to the grass-roots and impact teaching where it mattered, in the classroom. This was a powerful, and visionary move and had immediate results.

The establishment of the Foundation was a good idea and is still very effective. It helped focus public attention on the critical lack of geographic literacy, but it also plays an ongoing, monetary role in the provision of educational programs. It brings together the financial resources needed to fortify the situation and distributes funds where they can make a real difference - in the support of programs that improve and enhance geography education.

Where does all this money come from? First, the Society demonstrated its commitment to geography education with an outright gift of twenty million dollars to launch the foundation. Next, an additional twenty million was pledged to match contributions to the foundation through a challenge fund. Finally, it targets corporations, businesses, public policy-makers, Society members, and others, whose combined influence and financial resources, add impetus to the national drive for geographic literacy.

The Foundation extends grants for projects that meet the goals and standards of the Geography Education Program, especially undertakings that could have national impact, or those that could serve as demonstration projects for replication elsewhere. The plan is to make funding available for viable proposals that stimulate a greater understanding of geography in the educational arena.

The Geography Education Program established Geographic Alliances in many states as it was joined by educators and geographers in their mounting concern for geography illiteracy in schools across the nation. A Geographic Alliance is a grass-roots organization that brings together the content expertise of academic geographers and the classroom experience of teachers. Alliances are the energy centers from which spring local and state-wide efforts to improve the quality of geography education. The top funding priority of the Foundation is to serve the needs of the Geographic Alliances, providing operating and program support for them and stimulating state and private support by pledging to match up to fifty thousand dollars a year raised by each Alliance.

The state-based Geographic Alliances bring together the enthusiasm and expertise of academic geographers, classroom teachers, educational decision-makers, and private citizens to promote geography at the local level. At present, there are Alliances in approximately ninety per cent of the states, but the Society is working toward its ultimate goal - the establishment of Alliances in all fifty states. Alliances are led by coordinators who manage Alliance programs and serve as key sources of information about state activities.

As always, the need to train teachers, in both geographical content and method, was immediate and an old tried-and-tested routine was introduced once again. Valuable use was made of the summer vacation time to carry out intensive classes in relevant material. Today these summer geography institutes are supplemented by week-end and after-school workshops, and they are very popular with teachers in the Alliance Network, allowing for flexibility and an opportunity to increase proficiency in their subjects. Upon completion of a course, graduates are obligated to train other teachers in their districts and to work closely with members of their state alliances to improve geography education. In addition, Alliances develop classroom materials that are keyed to the local curriculum, and they coordinate geography awareness activities. Alliance efforts are making a difference; increasing numbers of schools and universities across our land now urge students to take geography as a requirement for graduation.

Since 1987, the program has spearheaded Geography Awareness Week, a congressionally proclaimed, nationwide celebration. In addition, the Program publishes a periodic newsletter, UPDATE, which keeps educators and decision-makers informed about the Program's activities and accomplishments. Each issue also contains geography lesson plans for classroom use.

The National Geography Bee increases awareness of geography, encourages teachers to include geography in their curricula, and sparks student interest in the subject. Students in grades four through eight are eligible for this entertaining and challenging test of knowledge spanning the broad discipline of geography. The National Geography Bee takes place in three stages, culminating in the national finals in Washington D.C.

The Program enlists the interest and support of governors, state legislators, and other state and local leaders. Many of these leaders have taught geography classes, have issued Geography Awareness Week proclamations, and have become vigorous advocates of the Society's efforts to improve geography education.

Today's computer technology has not gone unnoticed by the Society's curriculum planners, and a variety of creative and innovative programs and facilities have been developed for enhancing the instruction of geographical phenomena. Computer technology can greatly optimize the way educators teach and the way students learn. The National Geographic Kids Network is a revolutionary telecommunications system that enables thousands of students in schools here and abroad to compare the results of their science and geography investigations. Students collect and analyze data with the help of computer-generated maps and charts, and they share their results using modems and telephone lines.

GTV is a unique system that combines a personal computer with a videodisc player to provide a flexible set of classroom resources. The GTV videodiscs offers two hours of video in forty short segments, with more than fifteen hundred pictures and nearly two hundred maps. The accompanying software allows teachers and students to view the segments in any order, to rearrange any of the individual images to make their own presentations and to access the wealth of information in more than seventeen hundred captions.

Modern computer course-ware kits link geography and science, history and other curriculum areas using the power and scope of technologies like multi-media and hyper-text. In *The Weather Machine*, students use telecommunications to receive current weather data. The students then prepare their own forecasts by analyzing the data on computer-generated maps. Other course-ware is also available to aid in instruction and keep students conversant with the latest technology.

Modern technology, computers and telecommunications are a wonderful advantage for our children in today's class-rooms. Students have immense power and capability at their fingertips, to store, interchange and manipulate data, maps and numbers. But geographers should not forget the old adage:

Nature is at once the school and the schoolmaster

The cry of Socrates, and the plea of Pestalozzi, must not go unheard. The Geographer's true laboratory is in the open space—the outdoors!

The National Geographic Society, together with the Association of American Geographers have made gigantic strides in progress since the mid-eighties. Here's a brief summary of their achievements.

"They are revitalizing the teaching and learning of geography in our nation's K-12 classrooms. This is not the geography of lists of state capitals, rivers and mountain ranges, but the exciting, problem solving geography that is capturing the minds of thousands of teachers and students across the country. This geography is a powerful discipline, essential to understanding our global home.

> *Together with legislators, educators, and concerned citizens, we share a fundamental objective: that following generations receive the geographic training they will need to better comprehend themselves and their environment and to compete and excel in the world marketplace."*

And now, it is up to you! Are you a geographer? Are you going to finish this book, locate the unanswered question and uncover the solution? Are you going to join me on Puget Sound? More important, are you going to join me in the promotion of educational provisions for the children of tomorrow's world, today? I hope so. Only you can answer this.

Whether you go on to study more geography or whether you have already picked up enough doesn't really matter. But I pray for all our sakes that you gain sufficient global awareness to be able to make informed decisions about your business, your life-style and your election choices. We cannot drift aimlessly in time and space. We are the key to the future. The challenge is before us and we must embrace it to make meaningful change for the better. The first step to achieve these goals is obvious to me; we must restore geography to its rightful place as a vital component of a well-rounded education.

Chapter Four:

In The Beginning

The earth is round. You know that, and I know that. It is an easy concept to live with today. But this is the 'space age' and we are in a privileged position. You can see pictures of the planet earth on your TV and in the newspapers most weeks. Just think, if you went to your local library and asked for some information on the earth you'd need a wheel-barrow to bring all the books home. Today, if somebody said to you that the earth was flat, you'd laugh, and wonder where had this person been all these years.

This has not always been the case. In the not too distant past, life was much simpler. People knew that the horizon was the edge of the world. The horizon was as far as you could see, and if you traveled to the horizon, you would immediately fall off the earth. And that was that.

Naturally, many superstitions and fears were associated with the edge of the earth. One of the edges was the equator. It was commonly believed that you would burn up if you went south as far as this edge, because this was a time when superstition and religion were close bed-fellows. People sought solutions to the perennial questions that pertained to the origin of the universe and they accepted many strange answers.

Religious beliefs contributed to the facts as they were then known. Heaven was in the vast blue space above the earth, and hell was in the unseen, terrible place underneath the earth. It all made perfect sense.

Fear of the unknown and religious dogmas kept things this way for a long time and the old superstitions and fears persisted right up until modern times. It can take forever to overcome popular beliefs even if they are entirely wrong. This has been proven time and time again throughout the ages.

Since the dawn of civilization people have tried to understand and explain the world in which they lived. This was not always easy since new scientific discoveries very often were in direct conflict with 'common sense' and the accepted church doctrine of the day. It took a great deal of courage to stand up against the flowing currents of humanity and state a new theory or open up a new corridor of thinking.

Every geographer wants to know what the world is, where it came from, and where it is going. You also have great questions that need to be answered. There is a simple scientific method for gaining knowledge about our planet and it will help each new geographer grow in awareness and understanding as more and more doors are opened.

Learning geography is easy. There is no secret formula, no mystery, and there is a method that you can use. This method did not come into being overnight, but has been used by all explorers and scientists since time began. It is helpful to put yourself in the place of an early explorer, on a new planet, and use the only methods that are at hand. You must ask questions about the things that are around you. Be wary to accept second-hand opinions and handed-down free information without first assessing its value and correctness. The earth is round - or is it? See for yourself. You could take it from me, and believe it, or you can go out and prove it for yourself and know it.

Early explorer had to look. Then he had to see. This is called the 'Look and See' method of cause and effect. This may seem too obvious, or maybe too simple, so that you may say - yes, of course if you look, you see. That is not necessarily so. If two people look at the same object, they will invariably not see the same thing. They may see different faces or scenes. Do a simple experiment with an acquaintance.

Look and See - it really works.

Look at a picture and write down five things that you each see. Then compare your observations with your friend's. Are they identical? Probably not. This happens because everybody's views are colored by the previous experiences they have had in their lives. In order to 'look and see' you must observe, notate and deduce. Look at something, write down what it is you see and then figure out what that means. Tie it all together so that it makes sense. This is how man first proved that the earth was round.

Can you prove that the earth is round? Let's take a closer look at this simple task! Early man spent a great deal of time observing the sky at night and the sun by day attempting to put some order into the cosmic mystery that surrounded him. He had some basic questions about the universe that he was observing and trying to understand. Why are there phases to the moon? Why is the moon larger when it is low on the horizon? Are there any other moons? Why does the sun go south in winter and come back north next summer? You too, like early man, can carry out a great many observations, make a lot of notes and deduce some universal truths, or you can begin from where you are in time and space today. All you need do is read and comprehend how other people have done this before us.

That is the beauty of civilization - and you don't have to re-invent the wheel every day. Comfortable in your own home, you can begin to know and understand these same facts, at a new and higher level. It may have taken people with great minds, or great luck, or great timing, many years to discover the universe of knowledge and truth, but here it is now for you to take as your own. They did not expend this energy, expense and time simply for you, but it is still yours. And this is important, since it has been the great quest of man to understand the universe upon which he resides from time immemorial.

So you are part of it, and perhaps you too can discover more and contribute much to the grand quest. You don't have to do this today. Begin slowly, come to grips with your environment, learn about your planet and your fellow space travelers, and some day you might find yourself in the position of discovering some new truth that can benefit mankind. It is not too much to ask of yourself.

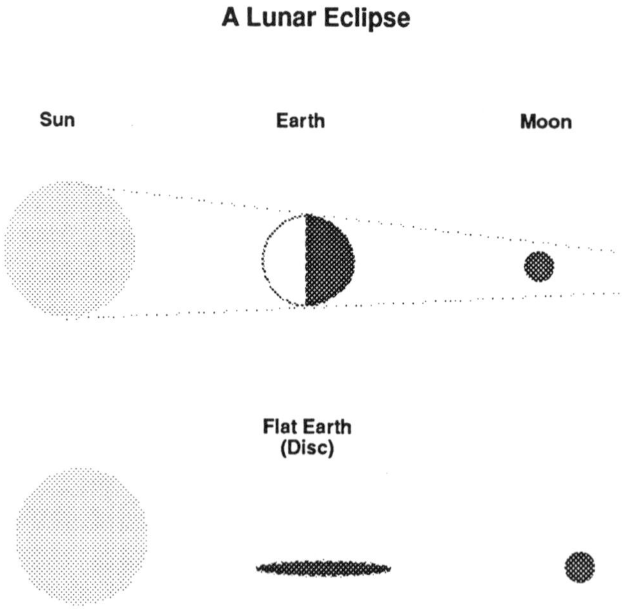

A Lunar Eclipse

Fig. 4.1 The Lunar Eclipse

The Greek Philosopher Aristotle was one of the first people to prove that the earth was not flat. The year was 340 BC. That was over two thousand years ago. He spent a great deal of his time observing the night sky and the stars because he was intrigued and fascinated by the moon and its phases. He wanted to understand its relationship with our planet earth. Every now and then a lunar eclipse occurred and he studied it. First he observed what happened. Each time he described in detail what he saw. Then he sat down and tried to deduce some common occurrences. From this he worked out a universal truth.

Aristotle observed that all eclipses of the moon were round. He realized that the lunar eclipse was occurring because the earth was coming between the sun and the Moon. But if the earth was flat why was it that during all the eclipses it cast a round shadow? At some point, he argued, there would have to be a shadow in the shape of a flat disc. But this never occurred. The lunar eclipse was always round. This could happen always, only if one set of rules applied. The eclipse would always have to occur at a time when the sun was directly under the center of the flat earth. This didn't make sense and didn't fit in with his observations. So he concluded that the earth was a sphere.

Aristotle continued to study the sky at night and he discovered that there was a star directly over the north pole. This was the North Star. He observed that it got lower in the sky as he moved toward the Equator. As he moved north it got higher in the sky. He deduced that at the North Pole, the North Star would be directly above the observer and at the equator it would be on the horizon. This, he concluded was another proof that the earth was round.

Have you ever stood at the edge of the ocean and looked out at the horizon, where the sea and the sky meet in a wide arc? What shape is it? It is easier to get a broad view when no tall buildings or trees are in the way. The horizon and the sky appear to be circular, for the amount you can see. Aristotle, too, stood at the edge of the ocean and observed the circular skyline. For him this was in compliance with his theory that the earth was a sphere.

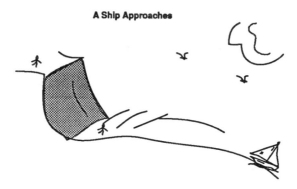

Fig. 4.2 A Ship Approaches

He noticed something else also. At that time ships had tall masts. He observed that the first part of the ship that would come into view as it sailed over the horizon was the top of the mast, long before the bow and the decks. It was as if the ship were sailing up an incline. He observed also that a person higher up on a hill could see the mast before a person on the beach. This could only happen if the ship was in fact sailing up an incline - or if the earth was round.

At night he observed the heavens and the stars. Every star and planet, moon and heavenly body appeared to be a sphere. Why then, he argued, should the earth be any different? Of course it shouldn't. Therefore, by association, it must also be round.

If the earth was round, as it was suggested by Aristotle, then it must have a diameter. If it had a diameter, it must also have a radius. These should be measurable. Mathematicians and astronomers began to focus on calculating the circumference of the earth. One of the earliest and the best known attempts to measure the distance around the earth was undertaken and completed by a Greek Mathematician, Eratosthenes, about 200

BC. This was a remarkable achievement and demonstrates clearly how observations, notations and deductions can bring answers. This is the essence of the 'Look and See' method. Here's how Eratosthenes made his world measuring calculations.

Alexandria was the center of learning for the ancient cultures, home to many great libraries. Eratosthenes was a librarian with access to a host of interesting views and information. He came across an account, written in an ancient papyrus scroll, that made him stop and think. He read that on the longest day of the year, the sun cast no shadow at noon, at a place called Cyene, so that the full glare from the sun penetrated down along the sides to the bottom of a deep well. That meant that the sun was directly overhead. At the same time, he learned elsewhere that the sun cast a long shadow on an obelisk in Alexandria, at noon, on this very day.

How could this be? "How could the sun behave differently at the same time, on the same day in two nearby places?" These observations would have gone unnoticed by the ordinary man, and of course they did. But Eratosthenes was no casual onlooker; he was a geographer and a mathematician and he needed to explain even the simplest of things so that they made perfect sense to him. He reasoned that if the earth was flat, this would not happen. But if the earth were round, not only would it cast a shadow, but the farther away you went from where the sun was vertical overhead, the longer would the shadow be.

This fascinated Eratosthenes, and he realized that he could calculate the circumference of the earth if he could measure the distance between Alexandria and Cyene. He hired a man to do just that. Can you imagine this early explorer, setting off on this adventure with his camel, measuring the stadia from Alexandria to Cyene. Have you any idea how many days, weeks, this journey took? At any rate, having measured the distance between the two cities, he postulated that this was equal to seven degrees of the full circle. How did he figure that out? He used Euclidean geometry to determine that the angle formed by two lines projected backwards to the center of the earth from the obelisks situated at each of the towns would intersect at about a seven degree angle.

He calculated the earth's circumference by taking the distance between Alexandria and Cyene and multiplying it by the number of times the angle at the earth's center is contained in 360. From the circumference he found the diameter to be about 7,800 miles. The correct Polar diameter is about 7,900 miles. Not bad for a man with only his brain, and a few camels, almost 2,200 years ago.

Eratosthenes Measures the Circumference of the Earth Circa. 200 BC

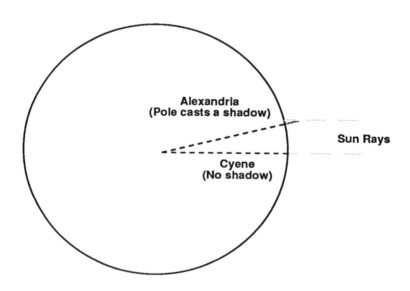

Fig. 4.3 Eratosthenes Measures the Circumference of the Earth

Today astronauts have been able to confirm the observations and theories that scientists have been postulating for a long time. Every mission into outer space is crammed with experiments and observations that verify theories that man has proposed using mathematics and physics over the years. From outside the earth scientists are able to photograph and observe from first-hand experience the curvature of the earth. In this way they concurred with Aristotle's findings. They were also able to verify and correct Eratosthenes' measurements.

Yes, the earth is round. You know it and I know it. That's the easy part. The point is, how did you come to that weighty conclusion. You didn't take somebody else's word for it. No. You went out and proved it for yourself. You proved it for yourself, using the same faculties for observation, recording and deducing as did the early philosophers and mathematicians. You used the method that has remained intact throughout the ages, and works for us today. This simple exercise is evidence of a new way for us - a way with geography.

Geography is fun; it is exciting, useful and it challenges the mind. Daily occurrences, things that were obvious and taken for granted, emerge under a new light as wondrous and not at all mundane. They serve to illustrate to us the intricate motions of nature as we are all involved in our great cosmic journey. This is our beginning - your's and mine. Here we start and prepare to explore the many facets of geography that promises to unfold for us, the amazing secrets of the universe. In the next chapter we will uncover some more ways that geography can be useful and challenging to us in our daily lives.

Chapter Five:

The Cosmic Glue

Aristotle and the Greeks proved that the earth was round nearly 2000 years ago, and now we grow up knowing and believing that they were right. Other actualities were determined by them also, things that were just as important to man in his everyday life. And there are still others that we do not accept today. We know better! Somewhere along the way geographers and scientists discredited them and proposed alternative explanations for the observed facts. Some were rejected out-of-hand, some lasted for a few generations, while others have withstood the test of time and are wholesome yet today.

As you stand in your garden or on your porch and look up at the night sky, you are observing the same sky that Aristotle, Eratosthenes and other great geographers studied. Can you see anything move? In a cloudless night you can peer into the heavens and see the moon and stars as far as the horizon, each bright pinpoint of light sparkling in the empty darkness that surrounds it. People have always asked the same questions that crowd your thoughts as you crane your neck skyward. Why can't I see the moon during the day? But wait, I see it some days and not others. Why? Where does it go? What is the difference between a star and a planet, between the sun and the moon? What does it all mean?

When you ask these questions, you are just like the first human to see and ask questions about the majesty and mystery of the universe. These are important questions and everyone should be able to answer them. The answers are not hard to track down, but you have to be prepared to do a little detective work.

As you continue to look, you may happen upon a shooting star or a meteorite, or a jet, or even a satellite, but nothing else appears to be moving. Yet, if you come back in a few hours you will notice that the moon is not in the same place. If you look very carefully, you will also notice that the stars are not in the same place, either. If you do this over a long period of time, just like the early astronomers, you will be able to distinguish patterns among the stars and soon you will be able to locate them, after a quick search of the night sky.

Can you tell what is happening? It is obvious that the moon and the stars are all rotating about you, as you are rooted on the planet earth. This is a good conclusion. Your observations are accurate and if you have spent many evenings taking notes of the celestial and terrestrial movements, everything is borne out again and again. You can confirm your findings by looking at the path of the sun by day. It rises every morning in the east and it sinks down to the horizon in a fiery ball every evening.

It is obvious that you are at the center and everything revolves about you. You could spend some time putting your findings down on paper, even mapping the entire cosmic picture. Perhaps you could make a model of the planetary movements to show as evidence, so that other people could see what you know.

Congratulations! You have now reached the same momentous conclusions that Ptolemy reached about the year 100 AD. Who? What was Ptolemy?

Ptolemy lived in ancient Greece, and he had a burning desire to know and understand the world in which he lived. He drew on the best observations made by others before him, but he emphasized the need for repeated, increasingly precise first-hand observation. He was tireless in his quest to put the cosmic puzzle together so that it made sense in the grand picture.

Ptolemy's observations and theories are preserved to this day in a thirteen part work that was so admired it became known as the Algamest - a word part Greek, part Arabic meaning the "Greatest". Ptolemy was the first to be the greatest, but he wasn't a boxer. He was a geographer. His Algamest was translated into a simple model known as an armillary sphere - an astronomical model with solid rings that displayed the relationships among the principal celestial circles. Let's go back to that year when Ptolemy introduced his Algamest.

What a breakthrough! Ptolemy's model of the planets, the sun and the moon, placed the earth at the center. That meant that the earth was stationary in the center of the universe and everything else revolved about it. His model was known as the *Ptolemaic Geocentric Theory of Cosmology.*

This theory was in harmony with the accepted views of the leaders of the Christian Church, since it seemed to present a perfect picture of the Universe. The earth was depicted as the most important planet because God placed man here. Beyond that, it showed vast open regions of space, at a suitable distance outside the planets, for Heaven and Hell.

They believed that God made the world, as is told in the story of creation. God was perfect, and his creation, the earth, was perfect. It made sense to believe that the planet earth was the center of the Universe. The earth was stationary and the sun, the moon, the other planets and all the stars revolved in circular orbits around the earth. Aristotle, a perfectionist, believed that the circular motion was the most perfect, so Ptolemy maintained that everything revolved in circles.

Ptolemy's *Geocentric Theory of Cosmology* remained the accepted "truth" for almost 2000 years. All this goes to show that second-hand, received knowledge can be misleading, no matter who gives it to you. Man holds on to some ideas that are very wrong for reasons that are not clear. It may have something to do with the way things have always been.

The terrible truth is, that your recent conclusions and Ptolemy's Model are both wrong - very wrong. Don't feel too bad, you're in awesome company. The Geocentric Theory is now recognized as one of the greatest elaborate mistakes in the history of mankind. Yet Ptolemy remained on center stage with his erroneous conclusions for over 1500 years. How could this have persisted for so long? At least your deductions didn't last that long. How could people have ignored the real picture for all those years? What were people doing? What were they thinking about?

Christianity replaced ancient Greek learning and wisdom - the advance of the one heralded the demise of the other. Christianity was accompanied by a blind belief in church doctrines that brought an abrupt end to a great era of incredible learning and discovery. Church doctrine and belief required acceptance of dogmas without question. The end of the Greek period meant the end of a time of learning - a fruitful time that produced individuals who were responsible for making vast strides in the understanding that we have for the Universe in which we live.

The story that accounts for the rise of a period of knowledge and discovery again is a fascinating one and every scholar should be aware of it. It is a constant source of inspiration and energy towards learning and freedom and a great way to instill your children with good attitudes and incentives for living their lives.

It was not until the renaissance period, in the twelfth century, that interest in the new world, exploration and the universe were again rekindled. By the 1100's, Ptolemy's Algamest was being studied by Arab geographers. Al-Idrisi, a very brilliant man, produced a world map in the year 1150, and in so doing, reintroduced Ptolemy into Europe. This was very important since it marked the end of the period of geographical darkness in

Europe. From that point on, geographers were on a new path to knowledge, rediscovering the quantitative tools that were their heritage from Greece and Rome.

Things didn't change overnight, however. It was not until the sixteenth century that people began to seriously question the theory of the earth as the center of the Universe, and it was well over a hundred years later that there was any general acceptance that Ptolemy was wrong - very much wrong. Nearing the end of the seventeenth century, it was finally concluded that the earth was not stationary in the middle and that Ptolemy had erred. This made way for a new and stimulating revolution in the study of the universe - a study that still persists today. But what of the great geographer - Ptolemy?

Ptolemy was rejected like an old shoe. Nobody wanted to associate with his views. What a great travesty of justice that Ptolemy's other scientific and intellectual achievements were totally overshadowed by the weight of such a colossal error, and today very few people realize his other accomplishments. But we can recount his one faux pas - the model of the cosmic plan with the earth at the center of the Universe. Yet today, in more ways than you may wish to imagine, we are influenced by that great geographer's work.

Ptolemy's work *Geography* contained a map of the world, which had profound influences on mankind. In the absence of first hand knowledge of the planet, he calculated the missing pieces - Terra Incognita. It was not perfect, to say the least. He exaggerated the extent of the land mass from Spain to China and underestimated the size of the ocean. This was a huge mistake and it had an earth shattering effect. That mistake encouraged a young adventurer to undertake a journey in the year 1492 - a journey into Terra Incognita, that according to Ptolemy's map, was not too distant. This man was, of course, Christopher Columbus, and the rest is ... geography.

The framework and the vocabulary of our maps are still shaped by him. In his *Geography*, which aimed to map the whole known world, he listed places systematically by latitude and longitude. This grid system is still the basis of all modern cartography. Every school student today learns to locate places using latitude and longitude, and we still orient maps the same way that Ptolemy established it - with the north at the top and the west at the left. He also pioneered the method of representing spherical surfaces on flat maps, using a simple conical projection to depict the known world of his day.

The man who reigned supreme as magnate of geography throughout the dark ages was to be eclipsed by the new revolutionaries, as a response to historical change. Ironically, the title of the work that was to superimpose his Algamest began with the Latin words - De Revolutionibus. New and exciting names were about to enter the pages of geographical discovery - Copernicus, Kepler, Galileo and others.

In the year 1514, a most unusual man used his observations to question some obvious faults in the daily calendar that was widely used. In so doing, he wedged open the door of discovery a tiny crack, a door that had been shut tight for so long. And once the door was opened, the light flooded in, so that it could never be closed again.

Nicholaus Copernicus was not an astronomer. Neither was he a mathematician. But, like you, he had learned to make accurate daily observations of the world around him, and he was stimulated by the challenge. He didn't invent any mind-blowing mathematical formulas. He did something more basic. He used his innate talents to "look and see". What would you have done? He saw that there were mistakes in the seasonal calendar, but he didn't accept it like everybody else. He boldly said that the underlying theory must be wrong and he looked for another one. A mere amateur in the world of science, he developed his theory as a hobby and never would have published it but for the encouragement and suggestions of his friends and relations. So what was his revolutionary discovery?

Ptolemy's predicted locations of the planets were not accurate. How could they be? All his theories were based on a premise that couldn't be more wrong. He had said that the earth did not move. For 1500 years man had lived by a calendar based on Ptolemy's fallacy. Naturally, over such a long period of time the summer solstice was off by nearly two weeks and other dates were equally out of sync with the actual observed occurrences.

Copernicus was not the only person to see that the seasons and the calendar were not in sync, but as you learned earlier, not everybody looking at the obvious sees it. Farmers knew the season for planting was Spring time, when new buds appeared on the trees; and the season for reaping was Autumn, when the crops were ripe and the leaves were about to change color. It did not matter if the calendar agreed with them or not. They knew better. After all, their very livelihoods depended on their ability to read the seasons and to grow their crops.

The average man about the business of everyday existence may have had doubts about the calendar, but he associated the inaccuracies with the general disorder of things. Many issues that he didn't agree with had to be accepted as part of the bitter pill of life. But Copernicus' great ability was in his realization that there was something wrong with the underlying theory that could produce such a calendar. He wasn't content to ignore the obvious and proceed as normal. He looked harder for an alternative theory, and when it wasn't forthcoming, he devised one himself.

In addition, the prosperity that developed during the renaissance demanded an accuracy in the calendar that was not equaled heretofore. New cities were growing rapidly, and commerce, over land and by sea, was expanding as technology was advancing. It became more and more essential to be able to rely on a calendar that predicted the seasons with a degree of ease and accuracy. Copernicus pointed to the obvious flaws in the Ptolemaic Calendar and sought an alternative theory to predict the movement of the celestial spheres. In his preface to *De*

Revolutionibus he noted the confusion about the seasonal year. He concluded that there must be something wrong with a theory that produced this calendar.

Scientists had devised labyrinthine and burdensome equations for upholding the Ptolemaic Model. Copernicus' genius was in figuring out that the key had to be simpler. Simplicity was the principle process of nature, because everything in the heavens should move like clockwork. Essentially, he didn't change Ptolemy's theories about either the shape or the motion of the heavenly bodies. He simply changed the center from the earth to the sun. All the planets still rotated in circular orbits, but the difference was that, now, the earth moved.

This was his revolutionary suggestion. The earth actually moved. This was astounding information. That the earth should move was contrary to all known "truths" of the day. Not the earth, but the sun was the center and the planets revolved about it. Earth was, therefore, just like the other planets. No wonder that Copernicus would be a lone voice for many decades, but he believed in himself and he believed his theory to be correct.

This was the birth of the Heliocentric Theory of the Universe, with the sun at the center. His major work *De Revolutionibus Orbium Caelestium* was published while he was on his death bed. There was still no confirming evidence that his was a better theory than Ptolemy's. But now, at least there was another theory. Copernicus had pushed the door further ajar, allowing young and inquiring scientists a breathtaking view into a fascinating new world.

Three years after Copernicus was buried, Tycho Brahe was born. A nobleman, he had the influence and the wherewithal to carry out repetitive and painstaking observations of the heavenly bodies. He devised ingenious methods for carrying out intricate observations and verifying the movements of the planets before the invention of the telescope.

Brahe was aware of Copernicus' heliocentric view of the universe, but he chose to ignore it since he was an ardent disciple of Ptolemy. However, in order to overcome the obvious flaws in that system he compromised and came up with his own system that was called the *Tychonic System*. He was adamant that the earth did not move and the sun moved about the earth, but now he maintained that the planets moved about the earth and the sun. Brahe spent his entire life collecting data from his observations. He was so sure of his theory that on his deathbed, he bequeathed his lifelong work to his assistant, Johannes Kepler, with the directive to go and prove his hypothesis.

Kepler set out to do just that. But things didn't work out in Brahe's favor. The more Kepler interpreted his master's observations, the more they began to point toward Copernicus. He made his most significant discoveries trying to find an orbit of the planet Mars to fit Brahe's observations. Since the time of Aristotle, the circle had been the accepted perfect orbit for heavenly bodies. He tried and tried to find a circle arrangement to agree with Brahe's observations. He configured every possible arrangement of circles to make the observations accommodate the predicted paths, but to no avail.

Change was in the air. Other observations were causing the walls of the ancient regime to crumble, laying to rest the Ptolemaic theory of the planetary vision. Galileo Gallilei was an Italian astronomer, and his careful observations confirmed, also, that Copernicus was correct, but there was something not quite right with the orbits. Like Kepler, he too found that the predicted orbits did not match the observed orbits. He studied longer and harder, seeking the solution. He made painstaking observations, and he filled copious manuscripts with annotated recordings, but technology was not ready for the breakthrough that would finally reveal the answers.

Nobody, prior to this, had questioned Aristotle's circular orbit of celestial movements. Kepler gradually came to the realization that perhaps the orbit might not be circular and resorted to an ellipse in his calculations. The ellipse worked wonderfully, and Kepler destroyed a belief that had persisted for two thousand years.

Kepler's momentous discovery was the elliptical orbit of the planets, because now, for the first time, all the predicted movements accommodated the observed motion. Flushed with success, and enthused with the energy of discovery, he went on to make gigantic advances in physics. He used Copernicus' revolutionary theory, Brahe's voluminous collection of data, and his own mathematical genius to formulate the three laws of planetary motion that were to perform an indispensable part of Isaac Newton's discovery of general gravitation a century later.

Then, in the year 1609 the telescope was invented and Galileo applied it to astronomical observations. He focussed on the planet Jupiter, and to his delight he observed that several small moons orbited around it. There it was - the final proof, that put to rest forever, Ptolemy's theory that everything had to orbit around the earth.

Chapter Six:

Towards a Unified Theory

What does the work of Copernicus, Kepler and Galileo mean to people today? Is Ptolemy a real person for us or a figment of someone's imagination? How much do you know about your planet? Have you taken the time to really dig down and find out how it all comes together? And the rest of us --- are we in a better position than were our predecessors who lived under the shadow of the church and superstition? Or have we, by asking all these questions, opened the dreaded, proverbial canister of worms? After all, the year 2000 is fast approaching, and we are far from the dark ages. Or are we? Sometimes I wonder!

It seems that the more we learn the more we discover we have to learn. Does that make sense...?

Recall what happened when NASA sent a space probe to the farthest reaches of the Solar System - our Solar System, the one in which we live. And bear in mind that we are but a tiny speck in the cosmic configuration. Look at what's out there beyond our solar system, stars, galaxies, and quasars, moving away from us at velocities approaching the speed of light. Moreover, they have been traveling at these speeds for what seems to us like eternity. It took the probe forever to get out to Saturn and beyond to Pluto, and when finally earth began receiving pictures back, pictures that should have produced the solutions to age old problems, everybody was aghast. Instead of answering the myriad of questions that had baffled our scientists for decades, the latest discoveries uncovered myriads of newer questions, each seeking its own solution.

The more we knew, the more we knew we didn't know!

When the space probe entered the outer rings of Saturn, I remember watching it on TV. A general notion of delight and achievement pervaded each of us, together with a mischievous sense, mingled with that inevitable subterranean emotion that longed for the imminent rock collision in a wayward asteroid belt. Nevertheless, we were each affected just as if we were there ourselves, even if we didn't have the slightest impact on, or the remotest connection with NASA's missions. We were all part of it. The lowliest earthling who couldn't distinguish DOS from O-rings felt proud to be involved vicariously in his living room - because we are part of the new generation of space probing scientists and gallant explorers - the modern day Columbuses.

We were breaking new ground, exploring beyond the realms of the imagination - going where no man ... you know what I mean! Soon that delight turned to frustration and awe. Instead of answering all the questions that needed answers, right now, the great voyage posited thousands of questions that weren't even thought about before this. Suddenly the catastrophic truth dawned on us. It wasn't as if we hadn't heard it before, but *"when I do, I understand"*. We were nothings, minuscule punies in the greater galactic pageant.

Certainly put us in _our_ place!

What did we do? Well, the next day we went about our daily tasks. We continued to fend for our families, raise our children and tend to our businesses. What else was there to do? Sometime later, and quite unconnected, a statistic was released that proved that American school-leavers were by far the worst at geographical skills in the civilized world.

So where does that leave us? Are we still in the dark ages, when, for 1500 years, people knew very little about their planet earth? And what they thought they knew was wrong. These people were our ancestors, yours and mine. And so it begs the question that you and I must ask. What about our descendants? Will they look back at the last few years of the twentieth century and say the same thing about us? Will they say that what we knew about our planet earth was not much and, anyway, it was wrong? Will they say that we were the generation that punctured the ozone layer, depleted the oxygen supply, denuded the forests and poisoned the water? Or will there be anybody to look back? It's very confusing!

The point is, we are here right now, alive and vigorous, and what we do will impact our own future and that of our planet. I'm so glad that you are taking the time to read this book. I do hope that it will motivate you into taking action to safeguard what we have today and help to keep this planet safe and life-sustaining for our offspring. If we don't do it, who will?

This should be only a beginning, a good foundation, from which you can influence other people and know that you are doing good by it. This way you will live a fulfilling life. Doesn't it make sense that if we build colossal houses on the banks of the scenic rivers and thereby, introduce pollution, then the river valleys are not going to be scenic any more, and a joy for everybody? I know I'm not the only person who sees things this way! It's not enough to flaunt mega amounts of money, with which we plant unsightly edifices in the most incredibly picturesque places, but we are also expected to accept it as normal and even laudable.

There is the constant struggle between the nature of man and man's ability to harness and control nature. Medicine is a good example. Because of advances in medicine, man lives longer and mortality rates are lower, but that means a more crowded planet.

Then there is the distribution of wealth. Have you ever wondered·why so few people seem to control the most of the wealth in the planet? Doesn't it seem ludicrous that half the world is starving, while the other half is over-producing and finding it necessary to store the excess produce in freezer ships and silos. There are thousands of people dying of starvation daily in various parts of Africa and Asia, while the new European Community continues to build up what they call: 'lakes' of milk and wine, 'mountains' of butter, ship-loads of lamb and beef. The distribution is all messed-up.

It's very confusing. The price of progress points to the conclusion that perhaps life is meaningless. Was man happier when he was in the state of nature, self sufficient and solitary? Or is he better off today, a civic minded social being, far removed from nature. One wonders if the achievement of happiness wasn't relinquished for the pursuit of safety and comfort! And at what price? Man no longer knows what he is, and why he is. Maybe he cannot be truly happy again until he achieves a return to nature!

Nature is seen as the raw material for man's progress, in his search for the ultimate something. But conversely, man is the ruination of nature. There is a justified conflict in all our heads today. On the one hand, we revere nature, and yet we know we must master nature. Every school kid is faced with the realistic simulation, involving the need to accommodate the rising population and the need to conserve the green and pristine countryside. This is the conflict between nature and society that Rousseau talked about over 100 years ago. Cities are too large, and yet they seem to be necessary to provide for civic man. We live in a throw-away society, even though recycling plays a great role in it. We are a society of toxic trains and garbage barges, of smog and commercialization. In our hour of greatest triumph over the laws of gravitation, time and space, we target the poorest and often the most idyllic areas of our land, such as Indian reservations, as the sites for domestic, industrial and even hazardous waste disposal units - our future happy dumping grounds.

People work very hard to exploit the natural resources to gain monetary acquisitions without regard for balances in nature or impacts on the planet. For this reason we are now faced with shortages in many of the plants and animals that used to be a plentiful and wondrous addition to the landscape. We have long since grown accustomed to the depletion and near extinction of animals and mammals, the natural inhabitants of the land, not to mention the ocean. And yet man in his untiring quest for ... who knows what, continues to ply his superior knowledge and talents in the exploitation and misuse of the earth's resources. Fossil fuels are being reduced at an alarming rate and the natural protectors of the environment, like the ozone layer and the world's supply of oxygen, for example, are constantly being besieged by the results of man's unthinking actions in the never ending attempt to harness and exploit the resources that occur naturally on the planet.

The only answer is an informed and knowledgeable populace who not only are in favor of, but also willing, to make the right decisions. And that, my friend, is where you and I come in.

Obviously, since you are reading this book right now, you are asking the right questions and that immediately puts you outside the realm of the blind believer. Remember the old adage:

When I hear, I forget
When I see, I remember
When I do, I understand

That is the key - doing is understanding. You can hear and see great amounts, but only when you have the personal experience of doing, will it all make sense. Ask yourself, how much do I hear and promptly forget? Look at the earlier problem. The earth is flat or it is round! That is only a concept, an idea that may or may not have any real meaning for you, since you are on the surface and, as it were, too much 'in the trees' to see the forest.

The earth is round. Somebody tells you this, and you store it in the back of your head somewhere. Later, you see the photographs from outer space and it comes back to you. But you have gone farther than this. You took the time to make some observations and calculations and physically prove it for yourself. Only then did you really understand.

Now put yourself in your children's place. You can tell them that the earth is round and think that's OK. I've done my job, I told them that the earth is round. But that doesn't mean that they have the slightest understanding of what that connotes. Watch them play with the globes. See them manipulate the concepts of day and night and the international date-line. Then you will understand how important it is for them to get the hands-on practical application that fosters proper understanding. Use the building blocks of learning. This is how Copernicus would have done it.

The universe continues to be explored in the same way that the dark continent of Africa was opened up in the last century. Many great scientists, whose names are now household words, dedicated their life's work to unfolding the secrets of nature and the keys to the universe. Great changes and scientific advances had taken place since the early years of the renaissance. There was more, a lot more, to the universe after Kepler and Galileo had made known their epic discoveries. These 'explorers' simply opened the floodgates through which was to flow the river of scientific discovery and through which we were to pass into the new age of space and time.

Ask any young scholars today and they will tell you about Galileo. More than likely they will inform you that he was a very old and blind man who invented the telescope and was later silenced by the church for spreading his heretical theories about the earth moving around the sun. Whoever invented the telescope might not agree with this description, but it is close and it does give our school children something to say about one of the all-time great minds. I wonder who did invent the telescope?

At any rate, Galileo was not always old, nor was he always blind - probably a result of looking directly at the sun through the telescope - and he was a truly remarkable man. Galileo was, after all, a product of his age and just as prone to believing the accepted 'truths' of the day as were any of his contemporaries. But, like you, he questioned things. He was not prepared to take everything at face value just because that's the way things always were. These were hard times. Think of this frail old man sitting in the snow outside the papal palace for three days, awaiting the dramatic audience that would put him under house arrest for the rest of his days. But his ideas were not frail, and they have withstood the test of time.

The last time I was in Italy, I visited Rome to view, at first hand, the words that Galileo was forced to say at his infamous inquisition, just over 300 years ago. His 'failing' and eventual undoing, was his steadfast defense of Copernicus' theory about the earth revolving about the sun. The celebrated words are there for everybody to read, original text in Latin, with the translation in the vernacular. And what an experience, to read the words of such a venerable genius as Galileo. He was heard to mutter under his breadth, as he walked away from his inquisitors:

"Still, it moves" - (meaning the earth).

In his blind, painful old-age he opened up vast worlds of fresh information and introduced difficult new questions with which mankind would come to wrestle.

In the next hundred years after Galileo, many of the fundamental secrets that govern nature were described, with the unfolding and explanation of new precepts, including the law of gravitation, the principles that apply to the speed of light, and the theory of relativity. A number of other isolated experiments and theories surfaced also, and each one edged us nearer and nearer to the final truths as they related to the structure of our universe. As technology and scientific progress gradually evolved, many doors were unlocked and man was getting a first glimpse at the very forces that were the cornerstones of our universe.

They did not all come about on the same day, nor were they discovered by the same individual. In most cases each new revelation was a finite field of study in itself, and constituted what we now call a partial theory in the quest for the origins of the universal truths. Today we are still striving, with the aid of these truths, to create a unified focused theory that would reveal the elemental building blocks of our very existence. In the next few pages, we will explore how the coming together and meshing of some partial theories resulted in our most preeminent scientific advancements.

Each of the developments is an awesome tale in its own right, but it was the cohesion and complexifying of these partial theories that has resulted in the permanent contemporary picture. I'm not saying that what we have today is the perfect answer, since clearly there are major gaps in the portfolio, and new questions keep raising their never-ending heads. But we have come a long way in our understanding of the immensity of the universe and the concepts of space and time. Sometimes I think we can even see things in the fourth dimension.

The story of the coalescence of the various partial theories involves such epic luminaries as Isaac Newton, Albert Einstein, Christian Doppler, Edwin Hubble and others. No discussion on the grand plan of the universe today would be complete without referring to the activities and achievements of at least these men. I would even go further and say that as young people seek heroes and goals to which to aspire, it would be good for them to read the life stories of these great men, to see where they came from, what they were thinking, and what they accomplished for mankind.

Ask any scholar to explain the law of gravity and you will immediately be greeted with Isaac Newton's apple story. Every young geographer knows that the spark of creativity came to him, suddenly, as he sat under the apple tree in his garden.

Maybe Newton was a marketing genius who saw the natural intrinsic value in apples! Isn't it interesting how the apple features in so many archetypes of our society. Adam and Eve, the forbidden fruit, Snow White, Isaac Newton, New York, Personal Computers, and an apple a day .... hey!

It is a wonderful story and it served its purpose well to popularize Newton and his great discoveries. Newton may have received a bump on the head from a falling apple, but it is more likely that, as he was in a contemplative mood in his garden, a falling apple focused his attention on the reasons why an apple falls to earth, and the moon, so many times greater, does not. It must have been one of those days when the moon was still visible in the clear blue sky.

Isaac Newton was greatly influenced by Kepler's, and indeed Galileo's work, since the essence of the gravitational laws was already predicted in theory, but it was left to the mathematical genius of Newton to prove the effect. He introduced his classic masterpiece, *Principia Mathematica*, in the year 1867, and his work has dominated the stage ever since. Consequently, Newton's name became synonymous with advancement in mathematics and physics.

Newton and Galileo were so different and lead such opposite lives, and yet each came to be an immense influence on us today. Galileo had discovered the underlying principles to the three laws of motion, but Newton proved them. Today, every school kid knows these three laws. The first is the law of inertia, where a body at rest will remain at rest until moved by a force. The second says that force is equal to mass times acceleration. And finally, every action exerts an equal and opposite reaction. But you knew that! These are the very building blocks of physics.

Newton was a popular hero but was despised by and among his peers for his abuse of power and privilege in the justification of his great ego. Galileo, on the other hand, had to suffer much pain and loneliness as a result of his discoveries and beliefs. The tales of their lives make fascinating reading and should be compulsory for every school goer, so that we can better know whence we came.

Newton is known as the father of gravity. He laid the foundations of calculus, extended the understanding of color and light, examined the mechanics of planetary motion, and derived the inverse square laws, an element crucial to his theory of universal gravitation. His law of universal gravitation states that any particle of matter in the universe attracts any other with a force varying directly as the product of the masses and inversely as the square of the distance between them. And that's it. For this reason the moon does not come crashing down to render us senseless as we sit in the apple orchard in pensive mood. So rest easy!

Newton's laws of gravitation were adequate to explain much of the planetary motion that governed our galactic bodies, but in the year 1905, a young German physicist, Albert Einstein, produced answers to problems that involved space and time. Einstein's famous equation $E=MC^2$ is on the tip of all schoolgoing tongues, but few understand the concepts involved therein. The truth is that Newton's gravitation theory holds just fine at speeds that are slow, but it is inaccurate at speeds approaching the speed of light. Since virtually nothing but light can approach that speed, it holds then that for everyday requirements, Newton's theory of gravitation will serve our purpose well.

Newton's name is synonymous with the theory of falling bodies and the motion of the earth. It is not as well-known that he was also an expert on the visible light spectrum, and in particular spectral analysis. When we speak of visible light, we are also making reference to another kind of light which must be invisible light. Science was coming to terms with new methods of 'seeing' that were not associated with the naked eye. White light, as we know light, is one of the fundamental constants in the universe, and is called visible light. Newton carried out extensive experiments to discover the properties of visible light, and his results have had momentous effects on the development of physics and the search for a theory of the universe. Newton was able to refract light through a simple prism and demonstrate that white light was made up of other colors. He found that the spectrum of visible light contains light in the colors violet, indigo, blue, green, yellow, orange and red. This was to have far reaching consequences in the study of the origins of the universe.

How does light travel? At what speed does it travel? These were questions on which Newton worked to unfold the secrets of the universal forces that comprised our planet's existence. It was nearly a hundred years later that another Englishman was able to demonstrate how light actually traveled, and to illuminate the intricacies that were hidden for so long. In the year 1865, James Clerk Maxwell brought together two partial theories that governed motion and matter to demonstrate that light traveled in waves. He unified the equations used to describe the forces of electricity and magnetism into what has become known as the electro-magnetic field, and he predicted that there could be wave-like disturbances in this field, and that these would travel at a fixed speed, like ripples in a pond.

This was a discovery of dramatic proportions -- that light traveled in waves. The frequency and the amplitude of wave theory could now be applied to the study of light. The wavelength of waves determines their structure, so that if the distance between one crest and the next is a meter or more, it is referred to as a radio wave. Shorter wavelengths, those of a few centimeters, are known as microwaves, and those with tiny wavelengths, ten thousandths of a centimeter, are known as infrared waves. Visible light, that is the light that we see, has a wavelength of between forty and eighty millionths of a centimeter. Even shorter wavelengths are known as ultraviolet, x-rays, and gamma rays. The theory of waves is now a crucial element of light study and part of every student's training. The problem is, we get so used to x-rays and microwaves in our everyday affairs, that we forget where it all comes from and what the significance of it all is.

In the year 1842, Christian Doppler discovered what came to be known as the Doppler effect in wave motion. He was an Austrian physicist who described the change in frequency of sound waves caused by the relative motion of the source of the waves and their observer. The same effect can happen with light or radio waves. But this is scientific stuff. How can we understand it and does it have any bearing on our everyday lives? Why, yes! We are familiar with the Doppler effect in our everyday lives.

The Doppler effect occurs everyday in our lives.

It is the simple story of the sound of a train or an ambulance approaching, and receding into the distance. The pitch gets higher as the vehicle comes nearer and it gets lower and changes sound as it disappears away from us. Notwithstanding the fact that the pitch is constant, to me or to you as the observer, it seems to change.

Any kid will mimic the high pitched whine of the train as it approaches and then tail off as it recedes into the distance. But what does all this mean? How does the sound, which is always constant as it is emitted from the source, appear to one observer to be high-pitched and to another to be low-pitched? The answer is simple, if you can visualize waves of sound being emitted by the train. When the train is approaching you, the waves are squished together in front and the result is a high pitch whine, but when the train has passed and is receding into the distance the waves are dragged out behind and the pitch gets lower until it eventually disappears. In fact, the wave lengths are either compressed for a high pitch or elongated for a low pitch.

So what does the Doppler effect have to do with the origins of the universe and the quest for the unified theory? True, the discovery of wave motion in light did not in itself have an immediate effect on the drive to learn about our origins, but when it was allied with another process remarkable discoveries took place. Recall that Newton refracted light through the prism and created the spectrum of visible light from red throughout the colors to violet. The spectrum of visible light breaks into several component parts, like a rainbow in the colors violet, indigo, blue, green, that have the shortest wave lengths and in the colors yellow, orange and red that have the longest wave lengths.

The Doppler effect was observed in light so that with infrared visibility we could detect motion causing the compression of wavelengths in front or extending them out behind, just like the train. When the wavelength was compressed, that is, the object was coming toward us there was a move toward the blue end of the spectrum.

Conversely, when an object was speeding away from us, there was a shift toward the red end of the spectrum. This is known as the **red-shift**, that is, the reddening of light sent out by an object that is moving away from an observer. This science of spectral analysis had profound effects on the way we were able to view and learn about our universe. Little by little over the years more and more information became readily available and as the jigsaw puzzle was coming together, it was hoped that all the pieces would coalesce to pinpoint an acceptable solution.

How did we get to this present position? Using Galileo's telescope, Newton's work on the spectrum of visible light, Maxwell's theories on the motion of electromagnetic disturbances, Doppler's studies on sound and light waves, and Einstein's relativity laws, scientists were ready to carry out exact and deliberate observations in the distant skies. Consequently, two incredible discoveries were unearthed.

This happened in the year 1929 and the American astronomer whose name is associated with the break-through is Edwin Powell Hubble. Hubble is a household name today but for very different reasons having to do with the ill-fated Hubble Telescope. That project turned out to be one of the greatest, most expensive, flops for NASA in the early nineties. But earlier in the century, Hubble was a name that commanded respect - a name that was synonymous with great scientific advances and revolutionary views of the planet earth.

Hubble discovered that there were other galaxies besides our Milky Way. He undertook the task of counting and classifying these galaxies and in studying them he made his second remarkable discovery. When he focused on the spectra of stars in far-away galaxies he detected that all the stars were red-shifted. In other words, they were speeding away from us. In addition, another startling revelation made itself known. It appeared that the amount of red-shift was greatest in the stars that were the farthest from us. What did this mean?

The galaxies newly discovered by Hubble, were receding from ours, and the farther away they were the faster they were receding. The implications of these observations were immense. The universe, long considered static, was expanding. This has been called the great intellectual revolution of the twentieth century, equal in stature to the over-turning of the Ptolemaic model of the universe in favor of Copernicus'. For the first time in the history of mankind, there was a realistic view of the universe. From there we were able to learn, using powerful telescopes, that besides our galaxy there are some hundred thousand million others, and each one containing some hundred thousand million stars.

Hubble made a quantum leap - an abrupt change in knowledge - into the world of Sputnik, Gemini, Apollo, and the Space Shuttle. He catapulted us into the world of black holes, quarks and quasars. Oops! Didn't mean to do that. What are black holes, quarks and quasars? This is the terminology of the new age of space walkers and computer gurus who create violent collisions using sub-atomic particles in massive devices called particle accelerators.

Black holes are thought to occur when a large star, like our sun, collapses inward and compresses under its own weight, so that its density and surface gravitational force increases greatly. The result is a region of space-time, whose gravity is so strong that nothing, not even light, can escape. Since the star's light cannot escape it becomes invisible to us, but we can detect its existence by the effect of its gravitational pull on other objects in space.

The quark is a strange word. It derives from a passage in an equally strange book - James Joyce's Finnegan's Wake - 'three quarks for Muster Mark' - meaning three quarts of Guinness for Mister Mark. Such are the vagaries of the scientific nomenclature that it is to this we refer today for one of the elements in the quest for the fundamental unit of matter.

A quasar is the most distant body yet detected in the universe and is an extremely bright star-like object at the center of a far-off galaxy. They are so far away that some are more than ten billion light-years from us. That means that the light we see today was emitted by the quasar billions of years ago. This is mind boggling. A light year is the distance traveled by light in one year, and light travels at a speed of one hundred and eighty-six thousand miles per second. Just imagine how many miles that is per minute, per hour, per day, per week, per month and finally, per year. Or better yet, ask your talented teenager to make the calculations - without a calculator!

Since the early sixties, astronomers have discovered cosmic objects known as quasars that exhibit longer red shifts than any of the remotest galaxies previously observed. The extremely larger red shifts of various quasars suggest that they are moving away from the earth at tremendous velocities approximately ninety percent the speed of light, and constitute some of the most distant objects in the universe.

The quest today is the same as it was in Galileo's day - to understand the universe in which we live. The methods are basically the same too - observation, notation and deduction. We attempt to unify the different partial theories, each that can go only so far on its own, to come up with the unified theory of the universe. From the minutest particle to the immensity of space, from quantum mechanics to relativity, we seek to unify all the partial theories to arrive at the quantum theory of the universe.

Chapter Seven:

You and Your World

Can you name the planets in our Solar System? Which one is the largest, which the smallest, the farthest away? It is a good idea to get a bird's-eye-view of the nine planets in our Sun's orbit in order to gain a serious perspective on where we are at. Look at our near neighbors and see why there are no obvious signs of life on their surfaces. We know that some are searing hot and others are veritable ice-balls; some are surrounded by poisonous gases and others are constantly in the throes of violent storms. Doesn't it make you wonder? How come the earth is the only planet that is situated at exactly the right distance from the Sun to sustain life? Isn't that an incredible miracle when viewed in the light of the totality of the universe, since we are but one tiny star amongst the millions of stars in our galaxy - the Milky Way - and there are millions of galaxies beyond our little corner of space. It is almost too vast to comprehend.

It is a clever learning experience to put yourself on a make belief spaceship and view the planetary ballet from beyond the farthest reaches of our galaxy. If you had the time and energy you could build a model of this space drama and really get to understand how the whole scenario works. It is not too difficult now that all the pieces of the jigsaw exist in your local library. While you are at the library it may be simpler for you to just view the many numerous and colorful graphic representations that are readily available. When you get the opportunity to visit a planetarium it makes the visit so much richer and such a learning experience. Yet again you can write to NASA and request photographs and other educational aids from their vast reservoir of exploratory information.

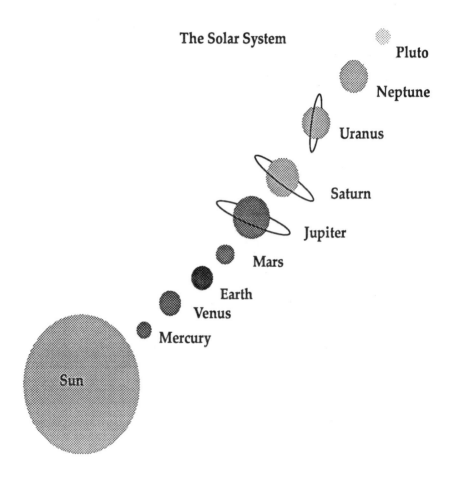

Fig. 7.1 The Solar System

I recommend that you play with the solar system in the comfort of your own living-room with your children. This can be an incredibly enriching educational experience for your children and for yourself since you will be learning and discovering together, new things about science, geography and about each of yourselves. I remember when I taught school and someone asked me what did I teach, I had to answer that I taught geography and I taught students. You can't have one without the other. In turn, we both gained from the experience since they learned from me and about me and I learned from them and about them. This can be even more rewarding when you and your own children share the learning, and the discovery, and the doing, and the fun, and the mistakes. It is not a difficult thing to do either, right there in your own home.

The following is a simple procedure to demonstrate the basic movements of the earth and shed some light on how night and day occurs. Even if they think they know it you'll be surprised how much they can still learn from the practical exploration. Of course you know that the earth rotates once on its own axis in twenty four hours and since that axis is some degrees off the perpendicular the length of the daylight varies depending on

Planet Earth

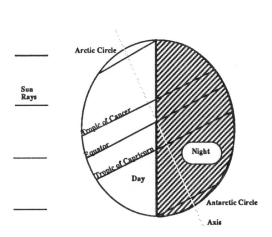

Fig 7.2 Day and Night

which hemisphere you are positioned. If you are on a part of the earth that is leaning toward the sun your days will be longer and the opposite occurs if you are leaning away from the sun. Then the earth orbits in a large ellipse around the sun once a year and these two movements together cause the seasons. Remember even if this seems too trivial and simple to you because you have known this *always*, the act of allowing hands-on direct application of the procedure to a young mind will be dramatic. When I hear, I forget - when I see, I remember - When I **do**, I understand. Be brave, make the first move today and resolve to carry out this simple experiment with a young mind.

The Seasons

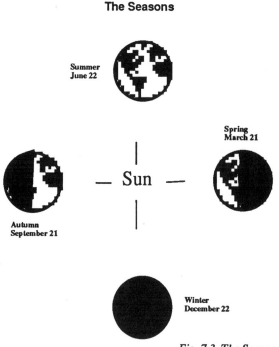

Fig. 7.3 The Seasons

Let the child explore with concrete hands-on experiences.

What do you need in order to carry out this simple task? Since you are going to do this experiment *in outer space* and you wish to view the earth as if from a spaceship it needs to be nighttime, or else you must get access to a dark-room. All you need then is a globe and a torch. Variations can be done, with good

Time Zones Across the United States

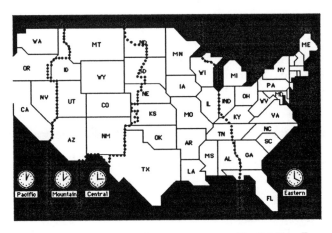

Fig. 7.4 Time Zones

effect, by using an overhead projector and a screen, or a slide projector and a dark slide with a pinhole as the sun source. It is a common fallacy that modern technology can solve all problems and that money is necessary in order to achieve education. This is a mistake. Many people believe that they cannot do any of this type of study without expensive instruments and telescopes, but they are wrong. Expensive equipment is not necessary; telescopes are not necessary. Begin with a globe and a torch and you will soon find that you are innovating and could possibly come up with many other ways that you could show effectively the motion of the planets and the universal picture.

The globe is the planet earth and when you pick it up you will immediately notice that you can turn it any way you like. The torch is the source of the light or the sun, and it too can be pointed anywhere. For best results, it is preferable that you shine the light with a concentrated beam on the earth near the equator. It is interesting to note that the distance and the orbit and the degree of slant creates a unique situation that has resulted in our particular environment and that could sustain life as we know it. Any other combination would have placed the planet too near the sun and we'd burn up or too far away and we'd freeze or in a strange orbit that would bring us too near and too far at different times so that we couldn't survive.

Now you need to create the situation where the sun is the center and the planet earth orbits around it in a large ellipse. To do this you will need somebody to hold the light so that the beam is directed on the globe at all times even when it moves in its large ellipse. Ideally, you might need one person to be the sun (hold the torch) and another person to be the planet earth (move around in an orbit). You (the observer) are just like the astronaut in space looking at the earth.

The first thing you might want to demonstrate is how the sun can shine on only half of the earth at any one time regardless of which half of the globe is facing towards the sun. It is easy to pick out day and night on the earth from your galactic vantage point and you could even observe twilight where the sun is not directly shining on that part of the earth, but there is an amount of light just the same.

Get someone (Mary) to be 'Monday morning' and get her to place her finger on the slowly rotating globe just as the dawn is happening. Let Mary follow through to noon and on into evening, twilight and then into Monday night, past midnight, and then approaching Tuesday morning. Now repeat that entire movement but place your home town at dawn on Monday morning. By the time you get to evening and night notice how somebody else, in some other part of the globe is still at Monday morning, half way around the world from you. Carry on until you reach Tuesday morning and notice that somebody else is still at Monday morning. So now you are a full day in front of that person and a half day in front of the first person. What an interesting discovery. Many years ago Vasco Da Gama, the first man to sail around the world, was very confused when he returned home to Portugal and he couldn't figure out where he had gained the day. He had kept a diligent diary and ship's log of all the events that occurred during his journey, and yet he was a day off.

Today, the international date line separates Monday from Tuesday in a more organized way. It is an imaginary line drawn from the North Pole to the South Pole that runs down the Pacific Ocean, from the Bering Sea between Alaska and Russia. Note how it is opposite the Greenwich meridian on the other side of the globe. Time zones and the International Date Line are important considerations in the fast moving world of today and people who travel a lot are forever adjusting their watches to synchronize their arrivals in different areas of the globe. You can buy watches that display different times in different parts of the world, corresponding with the various time-zones.

You can have a lot of fun learning about day and night and the international date line, but it is time to point out one of the peculiar arrangements of our planet that causes it to be so unique. The axis is twenty three and a half degrees off the perpendicular and this fact creates a situation where the length of day and night are the same only on two occasions in the year. These are called the equinoxes from the two Latin words for *equal* and *night*. Notice as the globe rounds off towards the poles that there is less and less daytime at one end and more and more at the other end. In fact, in one particular instance the sun never reaches the pole and at the opposite end the pole is in constant sunshine. This creates peculiar living arrangements for the Eskimos that live in the northern climes when they have midnight sun in the summer and no sun in the winter. With no light and no heat they have to adapt their living conditions to this exacting environment in order to survive.

Not only is the globe rotating on its own axis, an axis that is not truly perpendicular, but it is also orbiting about the sun in a large elliptical loop. Ask your *earth* to move through space around the sun, insuring that he/she holds the globe so that the axis always points in one direction. Prepare to stop him/her at four points in the ellipse equating each with the four seasons - Spring, Summer, Autumn, and Winter. First, decide if you want to be on the northern hemisphere or on the southern hemisphere. For the sake of the following procedure I am going to assume that we are on the northern hemisphere somewhere in America.

We will begin on March 22 at the Spring, or Vernal equinox. At this time the sun is directly over the equator and day and night are each twelve hours in duration. That means that every place on the equator will have twelve hours of daylight and twelve hours of night-time. Everywhere on the northern hemisphere will have shorter days and longer nights and the same will happen on the southern hemisphere. Now, as the *earth* moves further along the plane of its orbit and around the *sun*, we approach the next season - Summer. The sun is shining directly on a line of latitude that is twenty three and a half degrees north of the equator, marking the northernmost point that the sun can shine directly on the globe. This line is called the *Tropic of Cancer* and the day the sun is shining directly on it, is mid-summer's day - June 21. Notice that days are very much longer at this time in the northern hemisphere and from this day on the days will be getting shorter.

As the earth proceeds along on its yearly orbit we approach the next equinox - the Autumnal Equinox, when once again as in March, the sun is directly above the equator and the length of day and night is the same all along that line. But this time it is September 22nd. That means that March 21 and September 22 are similar in that the sun is directly over the equator for a brief time each year. The difference is that in March the sun is moving 'towards' us in the northern hemisphere, and in September 'away' from us.

Now as the year moves on the northern hemisphere leans away from the sun and the days get shorter and shorter until we reach, on December 22, the line in opposition to Cancer that is twenty three and a half degrees south of the equator. This is the *Tropic of Capricorn* and the day is the shortest day of the year, or the winter solstice. It heralds winter in the northern hemisphere and the corresponding summer in the southern hemisphere. Then as the year winds down the whole cycle starts over again. In this way we have four clearly defined seasons and four clearly defined observations of the sun between the tropics of Cancer and Capricorn.

My favorite season is Autumn because of the wonderful colors that display in the countryside at that time of the year. But I enjoy skiing in Winter and I love to swim in the Summer months when the lakes are amenable. Spring always reminds me of my youth, when I'm allowed to think that far back. Then everything seems young and energetic. I am delighted to live in a part of the world where there are four clearly defined seasons and each one has its own particular engaging attraction.

There are many advantages that accrue with exploring and experimenting with the globe in the manner just described. Not only will you, yourself, be reinforcing and verifying the ideas you have acquired about your home planet, but just think what it does for your children. Perhaps for the first time they will realize that, in fact, what you told them was indeed true - the earth is round. And more, they will be able to understand that the earth is not stationary but that it rotates about the sun, causing day and night. The tilt of the earth in relation to its planetary orbit will become, as near as possible, a real tangible concept offering reasonable explanations for the formation of the seasons. Only then will they be able to understand the significance of this tilt for us and it explains why so many races that reside on different parts of our earth have to adapt to such a wide variety of conditions ranging from arid through arctic.

This is a vital learning stage for all of us - doing and seeing. Only when we understand the rudimentary construction of our planet can we know how to care for it and preserve it in its pristine beauty for the future. We will continue to explore how you can learn about your world in the next chapter. At the same time you will see that it is not only fun and a great challenge to your mind, but you are developing geographic faculties that allow you gain an understanding of universal concepts and improve your retention of worldwide facts.

Chapter Eight:

What Can I Do?

Our house is a very, very, very fine house
with two maps in the hall
I used to be so small
Now everything is rosy 'cause I know

Today, I received in the mail a catchy brochure describing the school that my teenager attends. This brochure was very effective at promoting the staff and all their remarkable achievements, both scholastic and sporting. It was easy for me to conclude that it was a good school, a very good school by all accounts - excluding my bank account. In short, I should be well pleased that my son is attending such a venerable institution and innovative place of learning.

This is particularly true when I read about the incredible things that are going on in other establishments in the same city. It is hard for me to understand how schooling has degenerated to such a state, in recent years, where I am forced to say things like this. In my son's school there are no gang wars and no police in the corridors. I guess I should be satisfied.

I can't believe that I'm saying this. I should be pleased because my son goes to a school where the police do not patrol the corridors or conduct forced searches for weapons and drugs. Is it, I ask myself, too much to demand that a child should be allowed grow up in an atmosphere of peace and learning. Each of us parents must truly expect that our children will become caring young adults, fostered by an environment that is conducive to becoming just that - the people that we want them to be. Isn't this the least we can hope for?

The trouble with schools nowadays is that, too often, the teacher doesn't get a chance to teach - just enforce the rules and attempt to keep order. Great teachers can do both, but there are not an awful lot of great teachers and the others only succeed in maintaining the drive towards mediocrity which pervades our society today. Maybe that's the master plan! Who knows?

Generally, in any school there are 'good' teachers and 'not-so-good' teachers, and a few 'greats' - and that's about all one can expect. If you encounter one outstanding teacher in your educational path you should consider yourself very fortunate. We are all familiar with the individual that you can look back with affection at, in later years, and say; so-and-so really changed my outlook and my life, and left me with a burning desire for knowledge and learning.

Carpe diem!

In my son's school, however, it seemed from this brochure that not only were there a lot of very good teachers but there were also a lot of great teachers. And I wasn't disbelieving it. My son truly loves going to school. In fact, he can't leave the house fast enough in the mornings. At first I thought it was our house, but no.... it was obvious that he was learning vast amounts of useful and sundry knowledge in quiet content. And he comes home for dinner every evening - ravenous. So where's my problem?

I'm such a world fanatic that I always look for the geography in things. I love to look at maps and whenever I go to libraries and schools I want to see geography things. One of my favorite TV ads is one where this person is standing next to an almighty great globe. I don't even remember what the ad is about, but I can see that wonderful antique globe standing near the open window overlooking the verdant and splendid garden. Coffee beans, I think... from South America! David Niven... but I could be wrong.

I read the school brochure from cover to cover to know what goes on in my son's day. How does he get so hungry? There was a short description about each of the teachers that outlined their cognitive skills and the subjects they taught. Between them there was ample evidence of many fruitful years at Universities.

But yes... you've guessed it! Geography! No geography.

Not one of them had a degree in Geography, nor for that matter, any training in geography methodology - an absolutely necessary component in any geography teacher's kit-bag. Or if they had, it wasn't on the brochure!

I further discovered that in the entire brochure there was not even a single mention of the word geography. But hey, all the real subjects were described and the curriculum for each year was laid out in a visually agreeable format, for ease of reading. I was carefully escorted through math and science, through history and literature. All these concepts I understood, and I was also introduced to aesthetics and drama, music and sports and to new subjects that I hadn't heard before. But no geography! I re-read it, just in case I missed it, but no.

There was no geography! And there is more. This is only a symptom for what ails education in society today.

Recently, *Newsweek* did a Fall/Winter Special Edition on the issues facing education in the schools across the nation. It was a good series of articles and dealt with all the relevant items on the educational agenda, pushing all the hot buttons on the current social and economic arena. I read it with interest as did millions of other anxious parents who have teenagers in schools throughout the land.

But guess what was not even mentioned. What word did not appear in the one hundred or so pages that summed up the status of education in America today. Yes, you're very right - the missing word was geography. It is as if there is no geography going on in the country. This I find almost impossible to believe. After all we are living in the nineteen-nineties. And how does the country exist without geography? How do people get by from day-to-day without geography? How can they navigate their way through large cities and survive in a very sophisticated environment without this utility program? It perplexes me to no end.

Come to think of it, not one of my friends have any, what I would term *geography paraphernalia* in their homes. None of them have globes or atlases and I have never seen any maps or charts hanging on their walls. Is this pretty symptomatic of the standard of geography in the entire country? I think so! How about you? How would you rate in this picture. If I were to visit your home what geography items would I see, or is that a bad question?

How can our children grow up with any appreciation of geography, given the dismal picture that I've just presented? We have to begin somewhere and we must break this vicious circle of neglect. I can think of no better place to start than in our very domiciles - the homes wherein we learn most of our social values and the greatest portion of the attitudes that will remain with us throughout our lives. Here we can kindle the warmth of world knowledge and inspire global learning that will give our children a finer view of our planet and a greater understanding of the human condition.

What should you have in your house to make geography and global knowledge more accessible to yourself and your family? Where should you begin? I will promenade around some of, what I consider, the essentials for fostering a true love of geography and awakening a sense of exploration and discovery that will continue for life. I will outline my thoughts and books and gadgets that could help, and I will not always view the expenditure in terms of money, but often in terms that are far more valuable - those of time and commitment.

There are other matters to consider also, like the living-space and decor as it applies to your house. What works for me or for your neighbor may not, and probably won't, impress you in the least. And that's OK. You have to figure out what you want and how it would fit into your life. When you are selecting furnishings that will end up in your living room or your kitchen or study, the important thing to do is to try to achieve a balance between bareness, comfortable exposure and gross overkill. Sit down and think about the project first. If you have children, get them in on the planning, the selection and the preparation. Make it a fun and rewarding family experience. Simple things like this can be very effective, the most memorable and are often the least expensive. And what have you got to lose?

First of all, you need to decide some basic things. Are you going to place geographic learning items sporadically through-out the house, or are you going to assign one room - say, the study or the den - as the target for charts, maps and so on? Maybe your house has a library in it. Close your eyes. Imagine your dwelling place in your mind. Where would you like to see attractive books, colorful maps and splendid globes arranged in your home?

There are important considerations pertaining to location that have educational pros and cons, yet may have no impact on your decor plans and color schemes. I have visited bathrooms in people's houses that resembled galactic black-holes in outer space, with stars and planets that appeared to career wildly on the edge of eternity. I can only say that this motif created quite a unique mood in the bathtub. I also know a young couple, whose child's bedroom is decorated with skylight windows through the attic that seem to blend into infinity outside and clouds inside. Scary! I guess some people really go to town with the outer space theme.

You must be your own judge as to how far you want to go. In your home you must decide where the objects should be placed, hung, painted or sprayed. It is important not to lose sight of the reasons for embarking on this particular course. The objective is to create a warm learning atmosphere in your home that brings an element of geography into your own life and the lives of your family, so that they can appreciate their place in the cosmic order and better come to grips with, and begin to understand the mystery of life.

Globes serve a special dual purpose in your home. They are ideal geographical teaching tools, in addition to wonderful pieces of furniture. Studying a globe helps provide an understanding of basic geography and all children should be allowed ready access to one at all times. But look at what a globe would do for any room. It doesn't even have to be an antique globe to add color and distinction to your home.

Globes are normally found in good libraries and in some classrooms at school. In truth, they should be in every classroom and in all public buildings. I am a firm believer that each classroom in the school should have access to a globe for at least a short period of time during the school year. I would like to go further and say that:

> **Globes are a must. Every conscientious teacher should make sure that his/her classroom is equipped with a good globe.**

If the school cannot afford one, bring one in from home or buy one yourself. They are not that expensive and the advantages are far too numerous to overlook. Only then will you have a perfect plan of the earth right there in your presence for your children to look at and see our world.

No ifs, or buts - just do it! Put a globe in your child's classroom today!

Later, I will outline how to effectively encourage interest, by arranging classes in geography using a globe and some everyday utensils. Remember how learning takes place. It is very simple. There is no mystery! Encourage interest because interest is the key to motivation and motivation is the key to learning. Capitalize on natural youthful talents, like curiosity and learn how to focus their energy.

Interest is the key to motivation and motivation is the key to learning.

Why globes? Only a globe can give a correct picture of the earth as a whole. Because the surface of a globe is rounded like the earth's surface, the globe represents all parts of the earth's surface true to scale. Distances, areas, and directions can be observed without the distortion caused by projections used for flat maps.

Slow up. Let's look at this again.

Every flat map has something wrong with it. Either the distances are wrong or the directions are wrong or the area of the countries is wrong. This is why in some popular maps Greenland appears to be as big as Africa and the North Pole is massive. But a quick glance at a globe will show you the correct sizes for these places and you can be sure that the distances and directions are correct also. Obviously maps have advantages too, since they are flat and very useful for doing certain kinds of work, but you should be aware of their limitations and be able to check things on your globe.

For younger children a whole view is important so that they can gain a mental image of what the earth looks like. Later, when their assimilative faculties are further developed, you can let them piece bits of the world together to make the whole. But right now, it is good for them to see the big picture and to know that the earth is a sphere with land and sea. Don't take for granted that they already know this. How could they, unless someone has taken the time and the globe to show them?

The proportions and positions of the earth's landmasses, physical features and oceans as they occur in relation to each other, are seen on a globe exactly as they are on the earth. This is important. Looking at a globe is the next best thing to being in a spaceship flying by planet earth. You can see the land masses and the oceans, you can make out the upland areas from the river valleys, and you can locate different cities around the world with ease. In addition, it is obvious from the tilt in the globe's north/ south axis that certain parts of the earth are in a better position to receive heat and light from the sun as the planet pursues its plane of orbit in space. None of this is obvious from a map.

A great circle, the shortest distance between two points on the earth, can be measured directly on a globe. Of course, this will only have a bearing on you if you are making an international flight and you need to know the shortest route. On a globe it's easy. Just place a piece of string between your starting point and your destination. That is the shortest route - known as a geodesic. A straight line on a map may not be the shortest distance between two places unless it is a particular map constructed especially to calculate geodesics. But then the area is wrong!

Globes are usually mounted on a center axis to show how the earth rotates, but they may just as easily be placed in a cradle with no obvious attachment. There is no particular reason, save aesthetic and practical functionality, for doing so. Some people like to be able to show the angle of tilt as it applies to the earth so that it is easier to understand day and night, the international date line and the seasons.

With special accessories many different relationships between the earth and the sun can be demonstrated on a globe, including the length of daylight, time differences throughout the world, and satellite paths. I recommend that you begin with a fairly simple globe that is practical, but serves the basic functions of demonstrating the physical features of the earth and the tilt of the axis.

A globe that displays a map of the earth is known as a *Terrestrial* globe. This globe is usually mounted with the axis tilted twenty-three-and-a-half degrees from the vertical to help simulate the inclination of the earth, relative to the plane in which it orbits the sun. A globe that displays a map of the heavens, is known as a *Celestial* globe. When you feel that you are ready, you could progress to one of these that personifies more complex structures - a definite challenge to the latent astrophysicist in the family. Have fun!

The globe market is a veritable candy store where you will be faced with the decisions as to what model to buy and how big. These are very valid questions and I'm glad you are asking them. You must decide which globe would best suit your living conditions, your family's needs and your tastes. There are numerous kinds to choose from and each one has a value and a unique charm and personality to add to your home.

The most common globe is a useful table-top sphere about eighteen-inches high, and is an accurate representation of the political delimits of the world. These are changing so fast, of late, that it must be a veritable nightmare for the manufacturers to remain abreast of the major name changes and political alignments that are occurring globally. I read somewhere that Rand McNally were updating their databases daily, but were still holding off with the plate production, in case the political boundaries would change again before the next printing. I guess it's a good thing that we have computers today. Think of what a nightmare job this would be if we had to do it by hand. It's true;

The world is such a global village nowadays.

There are many different models to choose from and some globes have smoother, better 'action' than others. Some are designed to display physical features, land and oceans. Others display cultural or climatic zones, while others can be manipulated to show great circle routes and geodesics. More come on simple stands and others are cradled in complicated dishes that move subtly in many directions and for very specific reasons. You can have a lot of fun choosing one that would suit your home. You may even want more than one.

These globes are usually metal and plastic, are very friendly to children and withstand a great deal of hands-on rotation and seasonal exploration. They are colorful and pleasant and by choosing a political, physical, terrestrial or celestial model you can get a suitable color blend that would make a fine addition to any room. The good news is that these globes are very reasonably priced and can be obtained in most school supplies' stores. However, depending on how much you have or want to spend, there are some wonderful mail order shops that deal specifically in beautiful globes that I find irresistible.

The ultimate living room furniture is a large standing globe, with historic looking maps, and navigation charts set in a fine wood base. Mountains appear as raised relief and you can actually feel them, while land masses are represented in pretty pastel colors surrounded by attractive parchment oceans. The entire work is rounded off with colorful cartouches and compass roses, handsomely complemented by a natural, finished, solid oak stand and a brass-plated half-meridian ring. How could anyone resist this magnificent piece of work?

Of course you can spend a great deal of money on furniture of this ilk, if it is new and probably more if it is not. Such collectors' items are rare and very valuable, especially if the maps are in good condition and if the base is intact and still functions. Imagine how a globe like this would blend splendidly with your other antiques to enrich your decor.

Some globes can be illuminated and this introduces an added spectacular attraction to the piece as furniture, while it also makes it easier and more inviting to study. By turning the light switch, the continents and oceans are suddenly aglow with color and every detail is more brilliantly legible. Some of these illuminated globes are known as two-way. When the light is switched off, the physical relief is displayed and when the light is turned on, the political maps are highlighted. Pretty cool!

Illuminated globes are a colorful addition to any room.

There are a host of other features that come in the shape of globes and supply tasteful touches to your home. Some come complete with bookcase and atlas to make a perfect reference section in your study. Some people have globes that open at the equator to reveal a secret compartment that I have seen used to store valuables and goodies. There is a very popular executive desk accessory that is a neat antique globe mounted on a finished walnut plate and accompanied by an elegant pen holder. This makes a beautiful addition to anyone's desk and the perfect executive gift for the aspiring corporate leader.

I have many favorites and it really is not fair to let me free in stores that display globes. I must admit that there is one globe that appeals to me as the best looking, and yet the most modern - a globe that continues to fascinate me because of the marvels it can unlock as I gaze into it. This is the clear celestial globe - a see through model - that reproduces the earth and the stars, with all the constellations, their human and animal figures highlighted when it is illuminated. For me, this represents the best of old world magic beautifully blended with today's technology and it is so attractive that I cannot stray too far from it when I am near this store.

But I also like those antique globes with the three-legged base on casters. I guess I'll just have to have them both, or I'll go visit my local library where I know there is a great big globe.

Once, I was in Washington DC and I went to see the maps and globes at the Library of Congress. Now that's living. If you get a chance and have a lot of time to spend, don't pass this treat up.

When my boy was a baby, I bought an inflatable globe for his room and hung it up with the other mobiles. I won't say that he liked it better than the ducks or the colored rattles, but I certainly did. Today, he still has an inflatable globe in his room and it does add a fine decorative touch to the mess. His mother, in domestic desperation makes sure that each year she gives him a Christmas present that includes a large colorful calender that features mountains and rivers and other geographic scenes, to balance the hard rock and the heavy metal. And me... I'm going to wait until someone brings out a wall poster that portrays Bart Simpson saying - *'Globes are Geodesic Man.'* What more fitting tribute can the future portend?

New bumper sticker: Globes are Geodesic, man!

Now there's a good idea. Go on... admit it. Next time that a friend has a baby, in addition to getting the pink or blue cradle suit, why not get him/her an inflatable globe. Not only are you adding color and mobiles to the infant's environs, but you are also giving the little baby a wonderful start in life with the first globe. I'll bet that mom and dad will benefit from it also.

Globes are terrific representations of the earth as it really is. But there are some things that you simply cannot do with a globe and you certainly cannot carry one around with you for navigation purposes. Every medium has its own advantages and maps are ideally suited for ease of use, they are handy to pack and store and are filled with a wealth of detailed information that can be invaluable to you in many facets of your daily life. Sadly, it is seldom that one sees a map adorning the walls of children's rooms or studies. Such a wasted opportunity!

Maps have several advantages over globes since you can get a whole view of the world on a flat surface at a glance. Only half of the earth's surface can be seen at one time on a globe, and while they do give a great overview, most globes are too small to give much information about any one country or region. Other globes are too big and so cumbersome that they are inaccessible and awkward to move. Maps, on the other hand, are readily accessible and easy to manipulate into an advantageous position so that you can quickly get the information that you require.

Maps add color and an academic charm to your home. There is an immense variety to choose from and you should have little difficulty in blending one with your present decor. Next time that you are planning to change the complexion of a room or refurnish your house why don't you consider the value of a carefully selected wall map as an addition. Get that traveled international feel into your life.

Maps add an academic charm to your home.

A map is a graphic portrayal of all or part of the earth and is very important in the study of geography. Maps have lines, words, symbols and colors that show us the distribution and arrangement of features upon the earth's surface. They also use scales and grids to help us locate features and assimilate the information with the real world. Each feature is drawn in a reduced size so that it can be shown on paper. One inch on a map could show a distance of one hundred miles on the surface of the land.

Almost everyone uses a map at one time or another. In fact, they are so necessary in most large cities that it is a mistake to go anywhere without one. Maps are common everyday occurrences on subways and underground systems where the only method of knowing where you are is a mental representation. Maps help us travel from place to place and to understand the world around us. They help us plan vacation trips and follow news events in all parts of the world. Business men use maps to find good places to sell their products.

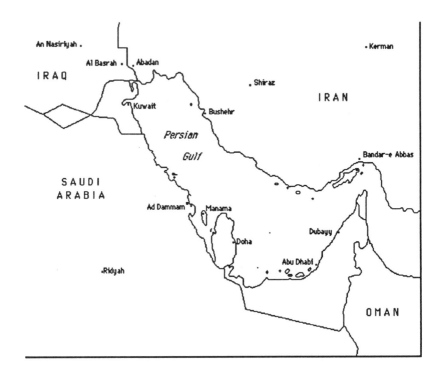

Fig. 8.1 Map of the Persian Gulf

Armed forces use maps to plan attack and defense strategies. The intensive use of maps was graphically brought before each of us during the war in the Gulf, when military personnel made their nightly television appearances flanked by very sophisticated looking maps. In truth, these maps were the same maps that we use in our everyday lives but they were focused on military targets and strategic maneuvers instead of bus routes and tourist traps. The physical features, sea and land configuration, major cities, and so on were overlaid with troop movements and targeted zones. Instead of the usual symbols that portray roads and the like we were shown bold arrows and graphic pictures of F-14's and aircraft carriers so that we could get a realistic picture of the war in an effective manner.

There are two kinds of map - a general reference map and a thematic map. The general reference map is just like the average road map, with which we are all familiar. It helps us to know in which direction we are going, which route to take, how far our journey is, and approximately how long it will take. A thematic map is used to emphasize a particular feature, like the amount of rainfall in a specific country or region.

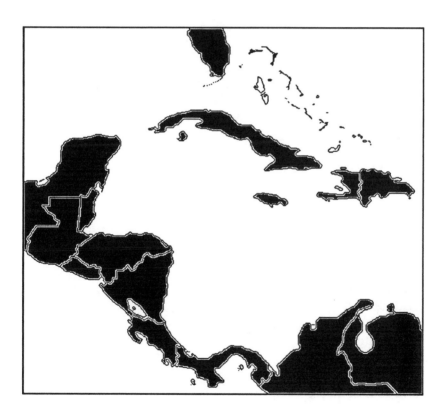

Fig. 8.2 Map of Central America

Special types of maps serve different uses. One type of map may show the number of people in every country in the world. Another may indicate the amount of rain that falls in different lands. Still another may show the different types of trees that grow in different parts of the world. There are nautical charts that give us information about coastal and marine areas and serve as guides to navigation and of course there are aeronautical maps that provide pilots and air navigators with a variety of information on surface features and air routes. Today, maps are extremely accurate since they are compiled with the latest in technological wizardry, ranging from infra-red imagery to earth observation satellites.

It is essential for young people to become familiar with maps so that they can gain a solid preparation for living in today's technological world. It's like riding a bicycle - once you learn you can never forget how it works. The skills that we learn at school will remain with us throughout our lives and this one skill, map-reading, will be useful again and again as we navigate through life on our planetary path.

Map-reading skills will be useful to us throughout our lives.

It is very important to have at least one map prominently displayed in your home. This could be either a map of the world, a map of the United States or whatever country in which you live. My preference is to have one of each so that you have ready access to your global village. And naturally, you should have a map of your local area for a quick and easy reference and to reinforce time, direction and distance.

Not all maps are the same, a cursory glance will highlight this fact for you. But rest assured, there are standard features and conditions laid down by international agencies that make them easier and more accessible for us to use. Thus, all maps are plotted with respect to the world standard geographic coordi-

nate system. Latitudes are measured north or south of the equator and longitudes are measured east or west of the prime
reference meridian of Greenwich, England.

All geographical positions are stated in degrees, minutes
and seconds and many symbols, scales and colors are also
standard throughout. There are sixty seconds in a minute, sixty
minutes in a degree, and there are three-hundred-and-sixty
degrees in a circle. If this is a difficult concept to grasp, simply
draw a circle and mark off the degrees in nineties and it will all
become obvious. Each ninety degrees is a right angle, and now
you've got the basics of any serious course in map-reading.

Maps are projections and since no map can correctly
represent any globe you must be aware of the limitations that
apply to each type of projection. Some maps will have distance
correct, others will have direction correct and others will have
area correct, but no map will have all three correct at the same
time.

Maps can be framed or plain pinned up on the wall. Either
way, the objective is to achieve a display of the world, so that
mental images are being formed. There is little use in having
maps neatly folded in the dark recesses of your desk drawer.
How many of us have been or still are guilty of this oversight?
Maps need to be seen.

Maps are great! If you have them - display them.

What's more, don't let the same display reside forever.
Change it every so often to focus on a selected theme. This is
particularly important in schools where young people are sub-
jected to so much imagery each day. Familiarity breeds contempt. Don't allow the imagination to stagnate because of lack of
stimulation.

One of the main advantages, and one of the best use of
maps, is the ability to teach location and establish a mental image
of what it is actually like at that location in the real world. Some

wall maps have photographs around the edges that depict the true landscape and living conditions that apply to that region of the earth. You can see realistic images of polar paraphernalia near the poles, tropical pictures near the equator and mountain scenes opposite the appropriate places and so on. This is very effective for creating mental images and in making the connection between the map representation and what is actually out there on the ground.

An even better way to bring maps to life in your home is to mark on your large wall map the towns and countries where you have lived or visited in the past. There are many methods to accomplish a colorful and attractive display that can have a deep personal meaning to the residents of the home. Place a colored spot sticker over the town where you were visiting, and opposite this on the margin add an attractive photograph, that typifies the region. Naturally, it should contain a colorful photograph with yourself resplendent in the foreground. In this way, you can become part of the map and get to own and feel proud of the display. It may surprise you how this could favorably impact your vacation time and bring energy and pleasure to other less remarkable periods of your year.

The project can become really exciting when a number of people in a family or a classroom share the same map. You could use a different color for each person and thus enhance ownership of a particular location. You are also inadvertently using some of the tools of map-reading to create this adventure, when you make use of colors and symbols.

I was told of one particular family that used to travel a lot because the father's job dictated it. At first, they saw this travel as an imposition, as a source of major disruption in family life and something distasteful. But, they were introduced to the idea of the travel-log wall-chart, and then this family, which now resides in Minneapolis, turned the experience into a grand-tour event in their lives. They expended great energy in setting up and maintaining their own wall-chart travel-log that brings hours of contentment and wonderful memories to the children. Each individual has a colored push-pin that is stuck in the chart to mark his/her personal and collective travel locations.

As the children grew older, the trips became diversified and differed from the parents quite a bit. Since they settled in Minneapolis they tied a colored thread from each distant location back "home" where the wall-chart is displayed. In this way, when you looked at the map, you immediately received the impression that this family traveled a great deal and that each member had traveled alone to different parts of the world. In fact, their real home was the world.

They are all very proud of their trips and they plan to carry on with more. This central wall-chart is always a major topic of conversation when visitors come to their house, and inquiring minds would like to know more about enchanting and faraway places. You don't have to be a great world traveler to successfully create this type of chart. It can be equally effective with a map of your country or even your county. But one thing is sure, once you begin one of these charts it will grow on you, and you could easily find yourself planning your vacations with exotic places in mind for your wall-chart.

A word of caution is necessary at this point. Many unsuspecting people started out slowly buying and displaying beautiful maps and charts, only to find that ten years later they are known by their reputation as 'map collectors' and globe connoisseurs. Are you going to let that happen to you?

An atlas is a collection of maps and charts. Every home should have a good one. Atlases are by far the best way to gain solid geographical information in a hurry. What should you look for in an atlas? Where should you keep it? Will you display it open at a particular new map each day? Or will it be placed at a good vantage point in your bookcase where it is readily accessible for all the family?

Atlases are wonderful. Don't go home without one!

There are a host of considerations that come to mind when one brings up the subject of reference works for general utility purposes. Again, you must decide what scenario or combination of ideas best suits your particular situation. You may not want your artistic two-year-old gaining access to a very expensive book with crayons in-tow. Or maybe you do! That's your decision.

One thing is sure, it will be used a lot, so it should have a good binding and be resilient to flat openings. For this reason you should probably choose a decent hard-back edition, for not only is an atlas a great source of information it is also a prime showpiece for your bookshelf. And whatever anybody tells you, there is a great sense of pride about owning a beautiful leather-bound, oversized atlas. I know many people who do not hide their showpiece in the bookcase, but have it displayed prominently on a stand of its own.

So you go out and buy a pretty expensive atlas. Now what do you do? Do you sit back and wait until one of the children arrives home from school some day announcing that he/she needs to look up some remote country in the *new* atlas. Suppose that never happens. Just suppose that the teacher never asks them about a distant land, or a war doesn't break out in some God-forsaken desert somewhere. What then? Will your atlas sit there on the bookshelf, still wrapped in its cellophane, keeping silent all its wonderful secrets? Yes! Of course it will. It is up to you to take it out and pore over it. You must open it up, let your fingers run over the outlines of the multitude of countries that are locked within its pages.

There is so much information stored in an atlas that it is usually difficult to know where to begin. Have you ever heard somebody respond to the question: "Read any good books lately?" with the answer: "Well, I'm actually stuck in my atlas now for a week and I find it fascinating, so colorful, so current!" And yet, if you open your atlas and page through it, you will find answers to almost all the questions you could ever have about your planet, the weather, scales, direction and a host of other things that you never even dreamed about. Do you know that people pay vast sums of money to learn these facts at universities? And here in your own home you have it all, in your beautiful new atlas. Dare I say; wake up!

Children love atlas games. There is no greater joy for them, than the exploration, learning and adventure of finding new places and learning new skills at the same time. The only thing that stops them playing atlas games is access to enough atlases and the fear of their parents and, sad to say some teachers, to get involved and immerse themselves in the latitude and longitude and in the degrees and hours and minutes. It is so easy to make education a cooperative quest for knowledge by enlisting the help of your children to learn about new places and you give them valuable vicarious experiences by discovering new and forgotten lands. You are also contributing to the training of the imagination and the emotions and preparing them for readiness, which is a prime requisite for learning.

Atlases usually have excellent indexes, and what better skill to teach your child in today's information-explosive world, than the skill of manipulating information by using an index with speed and alacrity. They can play in groups or alone depending on the circumstances, the age and the availability of atlases. Children love group activity and it is so good for them, helping them develop social skills and providing them with emotional involvement.

All you need do is locate a city or a river or a particular named feature on their atlas and give them a specified period of time to locate it using the index. A game even this simple will develop their imaginative faculties and sharpen their powers of observation. Try it:

"You have one minute to locate Timbuktoo, beginning now!"

I didn't know that Timbuktoo was in Africa. Did you? A game like this is sure to increase the vocabulary of new and exotic place-names that will flourish in your home and you can encourage it by association, with positive reinforcement. Help them learn in an atmosphere of fun and exploration.

Specifying a fixed period of time adds an element of urgency and excitement to the game and stimulates their motivation, allowing them to focus on the topic at hand. This is important for younger children when their span of attention is limited. They need to see a beginning, a middle and an end to the project. All this is good training for the enquiring mind. Avoid repeating questions or answers for your children. This destroys attention. And furthermore, you are only conditioning yourself to attend to their every whim. Instead, try this; pop snap questions at them, to develop their ability for instant recall, and sharpen their powers of observation and concentration. If you can help your children do that much, you are giving them an incredible start in life. You can stimulate the imaginative powers and the memory by posing a simple problem, as follows:

"What is the most westerly place in the United States?"

Naturally, everybody will have an opinion and everybody should hazard a guess. Do you know the answer? It's not important that you do. This is not a test. In fact, it might be better that you didn't. Even if you do know the answer, you should still verify that you are correct. Let them see the process; feel their way to the path of knowledge. You do not need to be a free source of information for your children, because even if today you have the answer to this simple question, there will come a day when you will not have the fast answer and they will not know what to do then. You can avoid placing them in that unpleasant situation by preparing them young.

The important thing is that the answer can be verified right there in the atlas and you can immediately confirm the learning experience. When they find out and verify for themselves the information becomes their own and an important learning block is cemented into place, as new knowledge is melded onto old knowledge in a colloidal mass.

This demonstrates that many people have opinions and they are sometimes right and sometimes wrong, but you can get at the correct answer by going to the source and seeing for yourself. What a breakthrough that would be and a neat skill to impart to your children for life. They need not accept reported second-hand information when they can go directly to the source to verify the truth for themselves.

In this way you can provide a wide variety of experiences actual and vicarious through observation and discovery. Notice that even though we are talking about geography and atlases you and your children are learning other things also - things that have to do with philosophy, with life and the meaning of existence. This is one of the ways that geography is such a synthetic subject and applies to many facets of your daily life. Let's look at some other ways that geography is intrusive in our lives.

The most common atlas is the road atlas, and every household should have one of these for planning and executing long trips overland. Vacation times can be wonderful times to put your children's new navigation skills to the test. It also serves the dual purpose of keeping them occupied and in touch with the trip when long stretches of driving are necessary.

How many parents are too familiar with the difficulty of sibling rivalry in the rear seat of the station wagon, that heralds unpleasant memories of the *big trip*. Careful planning and involvement can minimize the boredom that is necessitated by flat uninteresting terrain and rainy conditions.

Even if you know the route by heart and you do not need to consult a map, you should insist on letting everybody study the map so that they have a mental image of what is in store for them. Let them work out some answers. Where are you going, how far is it, what direction are you going in, and how long will it take to get there?

It is a good idea to appoint someone as the *navigator* for different legs of the journey and let him/her issue the instructions. Give them some important responsibilities, and be sure to follow these instructions. Don't just over-ride their decisions because you know more than they.

If there is going to be a problem with the route that is selected, stop the car and discuss the alternatives. Don't talk down to your children, talk across as equals. Remember you are all in this cooperative quest for knowledge and truth together and you are not the infinite source of knowledge. The journey is as important as the end. Point out what you think are viable routes, be willing to listen to arguments for and against, give everybody an opportunity to contribute, reach an agreement and continue the journey.

Look at this sequence again in slow motion. A typical disagreement occurs between driver and navigator. What do you do?

* Stop the car.
* Discuss the problem.
* Consult the map.
* Share some ideas.
* Share some refreshments.
* Talk across as equals to your children.
* Listen to alternatives.
* Be willing to try a new way.
* Allow everybody a chance to contribute.
* Reach an agreement.
* Continue the journey in comfort and happiness.

This teaches the children a lot more than just map-reading and interpreting the plan, and is a basic ingredient in the preparation for life. Build up their self esteem, make them feel good about themselves and let them know that they are making an important contribution to your trip. Be pleased and confident in the knowledge that your trip will unfold wonderful memories for you all in the future and that growing up with geography is useful and fun.

You can do an awful lot of positive things about the house to help your children become practical geographers, and sensitive users of our planet. And after all, we hope that they are going to undertake the care of our planet for the next generation. Geography makes a lot more sense when you view it in this respect - that we are making our minuscule contribution to the future of humanity by preparing these young people to accept the responsibility of the care of the earth for the next hundred years. Now your role is more critical and can have a vital impact on the people that will come after you.

There are wonderful geographical gifts that you can give to your family on special occasions, like birthdays, and especially as rewards for projects well done. In addition, there is a huge selection of geographical games and educational puzzles that could stimulate very creative fun with the entire family. Somebody has designed real catchy garments that are fashionable and geographic. The *Original Map Jacket* and the *Glowing Galaxy Jacket* are very splendid costume outfits and great casual wear for the outdoors. Have you seen the *Earth Scape* shower curtains? They're great. I love bathrooms that display colorful fronts in this manner. There are fabulous *Children's Illustrated Atlases* that help the enquiring mind make its way through the maze of information and scenery. Great idea for the pre-teen in your life. But for the Mom who has everything, how about a stunning *Earth Pendant*, a blue sphere with hand imprinted continents of gold.

You can choose from an innovative selection of T-shirts relating to Pollution, Acid Rain and other conservation issues, and there are Note Cubes, Note Cards and Mugs that are imprinted and colorfully decorated with highly detailed world maps. For the office wall, or any wall for that matter, there is an incredible *Satellite View of the World*. This spectacular image of earth from space is indeed a technological wonder. I think this is what we have been waiting for all these years from science and technology. This is the proof we needed. It is a color mosaic of satellite images virtually unobstructed by clouds, and the end result is a picture of exceptional clarity.

Yes, we've arrived at last!

Don't overlook your office, your work-space and your back garden when you are planning to bring geography into your life. I'm sure that as you are reading these pages you will have a lot more ideas that will suit your situation for achieving these lofty goals than I can unfold in these short few paragraphs.

You don't have to go overboard with the decor in the office to still achieve a positive image relating to the environment and geographic phenomena. There are many wonderful calenders that portray mountains and scenic valleys and they make beautiful additions to your office furniture. Don't forget the impact that a fine globe could make in your office, and even if your particular business doesn't have the least thing to do with geography you can be creative and still bring some elements of nature into the picture.

Other businesses lend themselves intimately to map and geography displays, especially where there is international and regional travel involved. You can create displays that resemble locations that your business touches, and collect postcards form foreign lands. You'd be surprised to find out how something simple like this could improve your client relationship and your customer's perception of you. Not to mention how a focus like that could improve your employees' attitudes towards your customers, who are not, now so far away.

In the long run, you are going to be the person to decide how best to include some geographic material into your present work environment. Give it some thought and try a simple beginning - then watch it grow. This could be the next great corporate breakthrough in customer relations! Who knows?

Everyone has access to a back-yard, no matter how small. Often, these places can be ideal sites, where you can start your children off in the spirit of scientific experimentation and inculcate a sense of discovery that will prepare them for later life. Start simply. Take something mundane that impacts us all and learn everything that is possible to know about it. Weather effects everybody in some way or another, but how many people can say that they are experts in predicting the weather. How many people go through their entire lives not knowing a cumulus cloud from a cirrus cloud, a fall in pressure from a fall of snow, or the cause and effect of each?

The back yard is a truly special place for geographers and I am going to dedicate a chapter specifically to the various kinds of activities that can take place there.

Our homes and work places are reflections of each of us - their owners; just look at the way pets, and especially dogs, reflect their owners. If you are a geographer it will be very evident in the selection of maps, globes and pictures that abound in your study and living rooms. The attitude that you exhibit concerning these items will be a positive influence on your children while they are in their formative years, and will help them develop sound educational values for life.

![black bar]

Chapter Nine:

The Room

Do you remember your days in school? Who was your favorite teacher? Your most despised and feared teacher? Were they not the happiest days of your life? Or is it better not to bring up the subject? Be that as it may, there is something that you can do today to insure that your children can have their best days at school and come away with a rewarding and pleasant experience.

This next section of the book will deal with the situation as it applies to the geography room in the typical school in this country. This chapter is for geography teachers, but also for school administrators, and especially for parents. I want you to pretend that you are a geography teacher and live this chapter so that you can be an effective proponent of geography teaching methodology. Remember that there are things that you can do to help your teacher and your children.

You are a critical component in this education game. Even if your sole achievement stems from the fact that you were able to point out to your child's teacher that s/he should attend one of the summer schools in geography teaching methodology run by the National Geographic Alliance network. Yes, you too can help!

But even if you have the best teacher in the world, there are other aspects of teaching geography that will impact the way your children perceive their education and can flavor their outlook. The physical environment where they receive their schooling matters and so too, do the daily surroundings where they are taught the initial rudiments of geography. For this reason I want to focus on the room where geography is going to be imparted to the young and energetic minds that fill our high schools. Please join me.

You are the teacher. This is your room where you must get to know your students and teach them about geography and about life.

Your geography room must be the most pleasant, lived-in room in the school not only for yourself but especially for your students. This cannot happen overnight, and you must apply concerted pressure on the school authorities to achieve your goals. Neither will it happen simply by filling it up with all the modern technological paraphernalia that can be used for teaching. If anything, these mod cons can be a serious source of distraction unless you know what they are for and how to use them to advantage. They can easily detract from your stated purpose which is to impart geographical skills, knowledge and competence to your students. The geography room should be a place where students learn, a place where they can feel welcome and safe in an inspiring atmosphere of erudition.

Let me begin by outlining what to avoid by way of establishing a base for your work. What you don't want is a dull, monotonous, drab room with poor lighting, a dingy, claustrophobic atmosphere and a sad teacher. You definitely want to avoid this type of setting. The last thing you want to convey to your students, is that geography is dull, drab, repetitive and claustrophobic. With that in mind you can now set out to design a work room, living/study room, and play room that is going to be your geography laboratory.

Where possible, establish one room in the school as the geography lab. Call it just that, and on the outside of the door print in great bold letters the following title:

The Geography Laboratory

Let there be no doubt in people's minds about the purpose of this room. This is where you are going to conduct your lessons that will impart the necessary skills and values that will turn out great geographers. It is equally important to establish this fact for the students as well as for the rest of the staff, who may not yet be prepared to grant geography its rightful place in the school curriculum. Very soon every student will know what goes on in this room and once you have set the train in motion it will carry on under its own impetus. The geography lab will be accepted just as the science lab and the language labs are intrinsic parts of the common school, today.

The effect will be cumulative and in a short time, you will find that geography will have a higher profile and more attention and facilities will be granted to it and to you. The establishment of a physical location will result in its prominent location in everybody's collective mind. Once you have established the room, it is vital that you do not send mixed messages about geography by teaching other subjects in that room.

You don't learn physics in the language lab and you wouldn't expect to learn French in the science lab, so don't allow the opportunity for other subjects to introduce retro-active inhibitions to your geography lab. You can do without that kind of attention, and your students do not need to associate the trauma of Shakespeare or the humor of Homer with your lab. Be prepared to wage this battle for a time until the last vestiges of social studies, as we know it today, have been eradicated from our school curriculum.

The moment you enter the geography lab you want to feel that this is a special room, and that good things happen in here. You want to feel welcome and warm and inspired by the layout and the atmosphere. With this in mind, choose a room that controls an uplifting view augmented by large windows. Many teachers shy away from rooms with large windows and distracting views, so you should have no trouble acquiring the kind of room you want.

Some teachers hesitate because they say that the pupils would constantly be looking at the view and would not be paying attention to their work, and let me assure you that, better they attend to that view rather than be bored by their work. If the geography class is boring then you are not teaching geography or you are using the wrong method. And if you are boring your class it doesn't matter what view or distraction they will seek, to avoid it. It is better that you have a view to compete with rather than lose-out to a fly crawling up the inside of a dull pane of glass that looks onto a dirty brick wall.

The key, of course, is to prepare and conduct classes that are anything but dull and conducive to staring blankly through windowpanes. It is better for the student to dream in verdant gardens than stifle in dull drudgery. When geography is taught to its peak, you yourself will look forward to the fine views, because you will have so little else to do but stare out to those lush gardens lest you interrupt your students who are working quietly and feverishly in the background. And all the while, you will have a class that is inspired and creative because they are in rich and pleasant surroundings.

If you have an option, choose a room that has a particular aspect that can be turned to your advantage in class preparation. I once had a geography lab that comprised the width of the top floor in the middle of the building, facing east-west with a door that lead up to a flat roof. It was the ideal geography lab with a great view from the rear over the playing fields and the river nearby, and a view to the front over the city and to the ridge where the airport was located.

From our vantage point we could observe the sun rising in the mornings and watch it sink to the horizon in the evenings, and track its progress throughout the year. On cloudless nights we set up our telescope on the roof and observed the moon, the stars and the planets. We spent many enthralling nights watching lunar eclipses into the small hours and we learned a great deal. I was sad to leave my geography lab when I moved on to another school and I never had one as ideal as that again. But you learn to adjust and make do, and use the facilities at your disposal to the best of your ability.

Paint the walls a bright, comfortable color and when that is not possible, cover up unsightly patches with colorful maps, charts and displays. I will never forget my dismay one September, when I returned to my lab from a European vacation only to discover that the principal had done a 'deal' with some past pupil who needed work, claiming that he was a painter.

The resulting art-work was, to say the least, stunning, when I walked through the door of my lab to be greeted by the ghastly, green color on the walls. It conveyed images of gut-wrenching bile in the aftermath of a hard Saturday night. The deed was done and I was forced to dig up all the maps, charts and photographs I could find, and proceed with the 'cover up' with colorful magazine pages, scarves and coats, anything, until such time that it was impossible to tell that my year had nearly been sabotaged by a not-so far-sighted pecuniary principal. I even enrolled the help of some of the pupils who lived nearby in rescuing the lab, and it turned out to be a great adventure and a remarkable recovery so early in the school year.

There should be some permanent fixtures in your lab, like a book case for your reference materials and a sink with a supply of water to carry out simple experiments and to help the plants grow. On one wall you could hang a permanent blackboard even if you have access to mobile boards. Chalk and talk is still a very valid media for instruction and one that you can use to skillful effect quite easily. If you know what you are doing on the blackboard, or greenboard, you can teach meaningful lessons and impart knowledge, skills and values.

I remember a blackboard technique that was used to outrageous effect by one of my most hated teachers, back when I was a boy. It was in a geography class when we were studying the physical features of France. Bona, the dreaded teacher, licked his forefinger and proceeded to make an outline map of France on the board with the wetted finger - no chalk. Naturally, the map disappeared as soon as the spit dried.

Pointing at me, he demanded that I 'put in' Paris. I knew what was going to happen, but I did as I was asked - it could have been anywhere, who could tell? Needless to relate, my attempt was the excuse he needed to unleash his wrath and I was soon beaten back to my proper position in life.

Next, he picked a 'good' boy, and having enquired as to the disposition of his father and mother he directed him to show me where Paris was. Well, I already knew where Paris was, but I kept my mouth shut. He confidently pointed to the same spot that I had pointed to and received the appropriate pat on the head as his reward. Bona rounded off my anxiety by reminding the class about the dangers of turning out like me and how much easier his life would be if all boys were as good as my 'friend' who was still smiling angelically.

Be that as it may, I grew up and thanks to the next teacher that came along I loved geography. Many years later I visited Paris and confirmed that I had put it in the correct position on that board so many years earlier, and to this day, I still have a great love for that city.

Blackboards are indeed very effective tools for teaching geography. I'm a great believer in tidy use of the board. Never mix concepts. All you need do first thing before you begin a lesson, is erase the old stuff from the board to get rid of any potential distractions. A *tabula rasa* will insure that there are no retro-active inhibitions remaining behind from a previous class or teacher.

Write neatly and use the space well. Don't write in a haphazard manner, diagonally, up/down and around corners, unless doing so constitutes a pertinent part of the lesson. Once you are finished with the concept or idea, clean the board. If you use the board sparingly, and introduce only one concept at a time, the blackboard will be an incredibly effective teaching tool. But, like all tools, if you are sloppy and inattentive to detail its effectiveness together with your teaching ability will be diminished.

There is nothing as bad as looking at a blackboard after school is over and seeing it overflowing with mixed concepts and ideas, bits of words and semi-erased phrases and a cornucopia of nonsense. We have all seen this in schools; a fraction of math in one corner superimposed on a drainage pattern and merging into some Spanish that abuts a scientific theory. Just imagine the effect of this splurge of information on a young, inquiring mind in its formative years. Used properly, however, the blackboard is a viable and profitable teaching tool and a valuable asset in your classroom.

There are a plethora of other boards that you can have in your lab also to supplement your class presentations. Each one has associated good and poor points and you must know these and carefully select the one that best suits your purposes.

As you set up your geography laboratory, you want to keep in mind that you are trying to create an atmosphere that is both easing and inspiring, pleasant to be in and uplifting to the spirit. There are two reasons for this. First, you want to create a pleasing location because your students will look forward to coming to your room and will have good associations and positive memories with the idea of the geography lab.

Secondly, this is the room where you will be spending a lot of your working day, and you want it to be a pleasant experience and a source of inspiration. You don't want to have to fight the room as well as the bureaucracy, and the discipline, and the weather, and the traffic, and so on, and so forth. Make it as easy as possible for your students to enjoy your geography classes. You are the most important ingredient in the formula for good education.

> **Even as they teach, men learn. So make your room work for you and in your favor.**

A fresh, innovative atmosphere can be achieved with a thoughtful arrangement of the student's desks, and the other classroom furniture. Each room will be different and the model that I draw may differ dramatically from yours. However, you can rest assured that the plan is good, if the end result is the same - a room that you and your students are proud of and one that fosters learning.

I like tables, rather than desks, because of the utility advantage of the flat surface when they are arranged together in groups of two or four a perfect work space is attained for conducting projects, group-work and detailed work on large maps. Every pupil should have his/her own desk and chair and you can anticipate their motivation and interest by training them to expect different types of class associated with a particular desk arrangement. For instance, if the desks are aligned in parallel lines it could signify a test, or if they are arranged in groups of four it might herald project activity, and if they are set up around the perimeter it could point to a debate or a field excursion and so on. You can use this technique to good advantage to motivate the students and to control retroactive inhibitions.

Where do you position your own desk? It is worth giving this question some serious thought. This could be the focal point for your teaching and for the students' sense of security. Wherever you decide to place it, stick with it. Make it a permanent fixture and your pupils will know where to expect to find you when they need you.

It is vital to build up respect and trust with your students so you don't have to spend your time policing them. If you begin by policing them, then you'll spend increasing amounts of time following up and chasing them rather than teaching them and they, in turn, will expend increasing amounts of time devising more sophisticated ways to elude you instead of learning. If there is trust and respect, this does not even come up as a topic of interest. You cannot demand respect, you must earn it and the way you set up your room and go about your business of teaching must win you that respect.

Your desk area is a wonderful opportunity for you to create a pleasant, friendly haven in the school and a place that the students will come to appreciate and enjoy as time goes by. You can make your desk the pivotal point of communication and learning so that the spirit of cooperation and trust is fostered and maintained.

This is the place to set your favorite globe and perhaps a 3-D map of your local area, with the raised features so that the students can touch and feel the mountain ranges and see the valleys. If their homes are marked on the map, it provides an added attraction and a perfect model for the instruction of maps.

Some teachers like to have raised platforms on which they place their desks, and I guess that's OK, if the teacher is short. But, I prefer to look across at my students and not look down on them. I think it is a prerequisite in the development of trust and respect. The classroom should be a center of learning, a cooperative quest for knowledge. Be fair and consistent in your dealings with your students and you will do wonders for their self esteem and for your own peace of mind.

Educare, from the Latin means to 'lead out' - it doesn't mean to fill up. You need not look on each student as an empty vessel to be 'topped up' with culture and knowledge; rather, each student must be provided with a suitable environment in which to grow and develop as he or she learns new and exciting knowledge. And your geography lab can be that place.

Every geography lab should have some fixed basic equipment to make class presentations lively and exciting. You can have daylight screens that are used with overhead projectors and other screens that are used in the dark. The overhead projector is a very practical teaching tool and it is often overlooked by people who have traditionally used the chalkboard. You can prepare large transparencies called overhead slides, which you can place on the projector to control the progress of the class and focus the attention on a particular aspect or theme. The main advantage of the overhead projector is that you can remain facing the classroom and know that the image is appearing on the screen behind your back. In this way, you do not have to turn your back on the class and risk losing their attention, or worse...

You can point to the slide on the projector and the image of the pointer appears in the appropriate place on the screen behind you. The screen is usually attached to the roof and can be easily pulled down during the class and re-wound for safe storage after use. You can make additions and corrections to the slide with the special overhead projector pens in all their resplendent colors, that will be faithfully displayed on the screen above and behind you.

I have witnessed some awful abuse of the overhead projector. I was once at a lecture where the maestro wasn't aware that he could point to the slide at his fingertips, and he had to jump to try to reach the appropriate spot on the screen. Naturally he could not reach the top of the screen and he was convinced that the whole set-up was a bad arrangement. The grand finale came when, intending to make a change to his notes, he turned his back on his audience and proceeded to ruin the screen with long lines and scribbles much to the horror of the class, which by this time, was too embarrassed to point out the folly of his ways. A-V 101!

It is essential to be able to turn your well-lighted room into a dark-room for slide shows and movies. I've been in 'dark' rooms and 'not-so-dark' rooms, and let me tell you that you are better off with none, rather than a shoddy attempt at darkness. You are only asking for trouble if you try to show slides that cannot be seen because the room is not dark enough or the slide is too dark.

Let me add that there is such a fantastic quantity of relevant material available nowadays, that no student should have to go through school without seeing pictures of volcanoes, glaciers, deserts, rivers and so on. Teaching physical geography can be so easy when you make good use of the available media.

Naturally, there is a whole lot of "do's and don'ts" that accompany dark-room presentations, but careful preparation will insure a perfect result every time. The last thing that you want to have happen is a prolonged period of no-action, while you are frantically fumbling in the dark in an attempt to get the slides into the projector - upside down and back-to-front. And hang on to the remote control, unless you can trust one of your students to change the slide at the appropriate time and keep it in focus as well. You don't want to show too many slides, just because you have them. Use the material to develop a point or to prove a theory and avoid overkill.

Teacher, know thy media.

Practice with each piece of equipment until you are comfortable with it, before you attempt to use it as a teaching technique in one of your lessons. Know how to load and rewind 16mm films (if you still have any) and this will expose your geography classes to a wealth of incredible footage on natural phenomena that you could never produce otherwise. If you have access to a VCR, it is always a good idea to have seen the movie before you discover some surprises with your class.

Know your school's audio-visual equipment so that you will be able to supplement your teaching with entertaining and colorful aids to good effect. I cannot stress enough that preparation is the key to bringing this type of lesson off without a hitch, especially when class periods are so short. It is also a good idea to know beforehand if the students will be exposed to the same type of instruction by another teacher before you. They may already be fatigued and burnt-out on audio-visual motifs before you even begin, a sure recipe for imminent failure and unavoidable disaster.

Build up your own collection of videos and slides that have worked for you over the years, so that class preparation is not always the impossible chore it can sometimes be. Be prepared to loan the videos out to your students to assist them at project work at home. You could have a reference library that contains books and videos and maps and so on. Let the material be accessible to your classes so that geography is portrayed for what it is - a living, useful subject.

Some teachers have computers in their geography labs today and are keeping their students in tune with state-of-the-art technological advances. There is a host of excellent bulletin boards and interactive software available from the National Geographic Society and elsewhere, that make geography on computers a fun experience and a meaningful learning activity. Once you get connected into the computing geography network it will grow exponentially and your students will be continually extended.

Remember that before your students can manipulate the software that is available they must have a basic knowledge of how computers operate and must feel competent to work at the keyboard. You may want to get together with the computer teacher so that the effort at preparing the students to undertake geography tasks on-line can be properly orchestrated and a good learning specter ensues.

There should be a solid and growing collection of maps in every geography lab. The ideal thing to do is build a map collection drawer-stack along one side-wall, where you can store large maps and charts without having to fold them and crease them. Your collection should contain black and white maps as well as color maps and should also contain geologic and meteorological charts. The maps should personify different scales and display regions from local, national and international places.

Use maps in every aspect of geography teaching and it is important to take care of them. Do not fold them, rather place them flat in the large drawers and they will last longer. Ask the students to only use pencils when they are working with the maps and to use care while leaning over them lest they get damaged.

It is important to instill a sense of respect for maps and atlases from the beginning and your lab will not have a bedraggled look after a term of use. In addition, you won't have to spend school funds in replacing equipment; funds that could more profitably be used to expand your presentation materials.

When your students develop a sense of ownership for the lab and its material they will be more likely to care for it and respect it. One of the most valuable skills you can impart to your students is the care and protection of their maps. Teach them how to unfurl and fold the everyday street maps that you can purchase at most gas stations and supermarkets.

Too often these maps are tattered not with use but as a result of mis-handling, are a source of frustration not information, and are of no value to anybody when the seams are illegible and there are large holes throughout. Have you ever witnessed somebody trying to pry a map into a readable position and then attempt to put it away "neatly"? There is a method and it can be taught to students - a very valuable skill for life.

Make sure that you have sufficient copies of each map to go around. That means that if there are twenty-four students in the class, pass out twenty four maps, one for each student. Otherwise pass out twelve maps, one for each couple working together. It always seems to work-out that the same kid ends up without a map each time there isn't enough to go around and that unfortunate kid grows up wondering - "why does it always happen to me?" You can avoid this with careful planning and by just being aware that there is a shy or lazy or a sleepy kid in your class.

Let them see and touch maps, noting the scale and the color schemes and recognizing the symbols. In a later chapter, I will expound more on the value of learning basic map skills for everyday living as well as gaining a broad preparation in a field that is fast becoming a major employer for the future.

It is very important to instill a sense of ownership for the geography lab. This can only be achieved by you with consistent effort and a clearly stated philosophy that governs the use of the lab. Each student and each class should somehow feel a part of the effort and should know that they have both individually and collectively contributed to its well being. You must work diligently to promote this kind of feeling and your job will be easier in the long run since the students will be pulling with you and not against you.

There are many ways to achieve this and often the methods that would work well for you might not be allowed in another school. But continue to get the students involved and to foster a team spirit around your lab and you will find that not only are the students pleased with themselves and their lab but you are also gaining important kudos for geography in the school curriculum along the way.

A good way to begin is by building something of value for the lab as a project with your class. Every geography lab can benefit from a good tracing table, and they are very inexpensive and easy to build. I recently received a letter from a teacher in Florida and her class letting me know about their fun project. In the letter she outlined how they had added a new piece of practical furniture to their geography lab. It was a tracing table... and then some. They started out to make a simple tracing table, but they added a display counter for their collection of rare minerals and erratics (exotic rocks). Read-on and discover what happened.

Basically, they understood that a tracing table was a broad flat translucent surface where you could trace a map onto blank paper. (The same effect can be achieved with a little more difficulty, by placing the map against a clear window by day. The light from outside will shine through and you can trace your outline on to blank paper. This technique is not recommended from the top story or against panes of glass that could shatter under pressure of a tracing pen.) Their initial design called for a plain table, with four legs and a flat top. Their plan was to add a six-inch wall all round, and finish it off with two strip lights in the box and a clear sheet of glass on top. Simple!

Then, flushed with the throes of victory and accomplishment, one of the students pointed out that they had just invented a 'cool' place to display the rock and mineral collection that was locked away in the teacher's drawer. And so, they set about arranging the minerals and rocks, each with a label and descriptive note and placed each neatly on the floor of the tracing table. It turned out that there was a geologic piece from many places in the United States and from around the world.

There was some petrified forest from New Mexico, some iron pyrites (fool's gold) from Oregon, some galena from Spain, colorful Carrara marble from Italy, some granite from a local quarry, white chalk from England, grey pumice and volcanic ash from Mt. St. Helens, and some beautiful finger-corral from Mombasa.

There was more, much more. But the best part of this entire story was how the students took ownership of the project, the rocks and the idea. It wasn't long until a pupil returned from a trip from another state and remembered to bring back a souvenir for the geology collection in the lab. You know what happened. Nothing promotes success, like success.

Soon there were too many samples to fit into the tracing table and the teacher had to make an arrangement whereby, every other month a different display was prepared. Then the students could look with pride on their contribution and share the wonderful memories that they brought back from their trips and vacations. Geography had a real-life purpose for them and it was fun, exciting, and uplifting. That table never needed a lock, it was never interfered with, and served as a true example of living geography and student involvement in their school.

The same thing applies to wall displays. I already mentioned the advantages of colorful displays on the walls, the need to focus on a theme, and the plan to change the display periodically to reflect the new focus. A good idea is to dedicate a section of your display to daily issues that can be updated every morning with current news events pertaining to the school and the class. This gives the students a sense of belonging and involvement. You should also have a national and an international section. Once a month, you could display physical, political, climatic maps, charts, pictures of people and scenery. The objective is to allow your students grow familiar with the environment and living-conditions of people in other parts of the world.

You will want to preserve a portion of your display area to focus on the particular topic that you are teaching this month, or semester. This maintains a formal image in front of the pupils and assists you in class motivation and preparation.

You must decide yourself what and how much display will best suit your own and your class needs. But one thing is sure - displays work. They work in brightening up a classroom, they work in motivating the pupils and in promoting interest in the subject.

Displays work! Colorful displays work colorfully.
Displays brighten your room.
Displays motivate your students.
Displays reward your students' efforts.
Displays promote interest and curiosity.

Finally, there should be a portion set aside to display the results of the students work, events and projects. This is the vital link that fosters self esteem and a positive sense of achievement, that leads to motivation, interest, and learning and that builds on the respect and trust that is so effective in a smoothly run class. Display their work liberally and with much praise.

Always praise their attempts even if you know that they could do more. Then help them to do more. But let them associate good work with praise and not with punishment. Show all pupils equally. Do not fall victim to the fallacy that if you display only the best projects that everybody will see how good your class is. Be fair and proud of all your students equally and they will respond a thousand-fold to your attention.

Show their projects, their photographs, their essays, their prizes and their test results. Show anything that will make them look like the kids they are. In this way, you can be assured that your walls will be living evidence of your teaching brilliance and proof that your students are happy campers.

I will never forget a fateful project that developed one year in my geography lab. We had been working on 'Rivers' for a week, but the truth was that I was up to my proverbial waist in alligators because there were just too many concepts and abstract ideas for a class that wasn't all that bright, just energetic and inquisitive. There was a major river running through our city, but it was in its old-age stage and was sluggish and smelly and plumb dangerous and I was not prepared to risk the consequences of taking this class of fourteen year-olds to the banks of such a trap.

That left no alternative or so it seemed at first, but to resort to slides and books and diagrams on the board. But I had learned that I should use the things that are geography in the teaching of geography and I wanted to avoid using second hand substitutions where-ever possible. Actually, the class was just not responding to the text we were using and the slides were not that good. It was a very difficult class to motivate.

Use the things that are of geography, in the teaching of geography.

I knew that I could arouse their interest if I could let them see a young river valley, with interlocking spurs, rapids, and braided streams, but where was I going to get the likes of that at short notice?

So I resolved to get the class to build their own river in a sand box, right there in the lab. At first they were disbelieving and thought the project impossible. In truth, I did too. But I certainly had their attention and the construction of the river was the talking point of most of their conversations for the next week, a fact attested by both the woodwork and the science teacher who were smiling, tongue in cheek, at my naivete.

The first thing we did was build a waterproof box in the dimensions that suited our lab. It turned out to be a pretty big box - eight foot long, by three feet wide and a foot deep. It looked like an enormous coffin and the word was out all over the school, that something devious was going on in the geography lab. Rumor had it that we were about to bury the French teacher. Next, we filled it up three-quarters full with sand from the nearest beach, which was twenty miles away. Do you know how many bags of sand a box that size can hold and can you imagine how heavy each soggy bag can be? It definitely got their attention as we hauled them up the three flights of stairs to the lab!

Sand Box

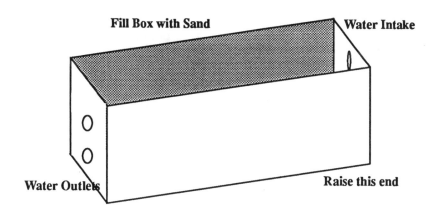

Fill Box with Sand   **Water Intake**

Water Outlets   **Raise this end**

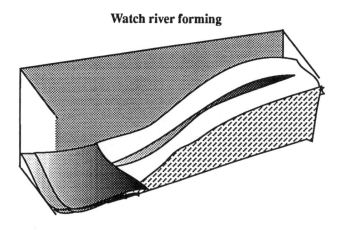

Watch river forming

Fig. 9.1 The River - Sand Tray

We needed a supply of water to start our river and we needed an estuary to drain our river - two pretty obvious items when you are dealing with rivers. So we set the 'river' up high enough to drain into the sink, and that was good because it was now at about the right level for a viewer to stand there and look in over the side at the action. We attached a hose at the mountain side of the box and we drilled a hole about eight inches up the side of the ocean so that our river wouldn't overflow onto the floor and down to the French lab below.... even though some of the students thought that this would be a super idea. But, I knew that Madame de Stael wouldn't be overly impressed and I wasn't about to find out.

Finally, the big day arrived and everything was in place. All the sand was built up in picturesque mountain formations and 'rain' was forecast by our meteorological division for about two o'clock in the afternoon. All the students were excited and the dignitaries (including the science teacher and the woodwork teacher) having arrived, the 'river' was turned on.

What a disappointment! I should have known. Nothing happened. There was over forty anxious people, including my principal, Madame de Stael and the other teachers, staring fixedly at the water disappearing into the sand. Eventually the 'estuary' filled up with ugly, soggy sand and began to pour out, sand and all, down the sink hole. I was mortified. The principal leaned over and turned off the water to save the plumbing bill no doubt, since the sand was already clogging the drain. Then he left quietly - as only principals can.

Back to the drawing board. Why was our river not working; did we have the essential ingredients in their correct places? The 'river' continued to be the centerpiece of the lunch-time conversation, much to my embarrassment. I learned an important lesson from all this - I learned never to invite anybody to witness an unveiling again.

Each day I tried something different and each day the same result would transpire - a wet, soggy mess on the floor of the box. Then one day we had a visitor to the school, an old inspector, who was monitoring the progress of some young teachers. He too was told about the 'great fiasco' in the geography lab and he was immediately captivated by the experiment. It turned out that he had been a geography teacher in his day, and had often thought about generating his own river. He spent all day watching and 'playing' with the flow and finally he came to me and said. "I think we need to raise up the mountain end, so that there is a real slope for the river to develop." Remarking the use of the royal "we" I thought that this novel approach was as good as the other 'novel approaches' we had already tried. Then I realized that the thing weighed nearly a ton, especially since the sand had become wet somehow.

There was nothing for it, but to procure some plastic bags and a shovel and begin again. That day after school, the inspector, who had just about taken up residence in my lab, about half the class and I donned our grubbies, emptied the box and raised up one end to a height of about six inches. Then we poured the sand back in and recreated our mountains. We were ready to try it again.

The result was dramatic; even better than any of us had imagined and now I think even more effective than if it had worked from the very beginning. Naturally, it became the highlight of the school year. As we watched we saw developing before our eyes a river valley take shape in its youthful stage, then move down to the lowlands and spill into a beautiful estuary through a perfectly shaped alluvial fan. It was incredible and everybody was awestruck by the unfolding features. We left it flow overnight and in the morning there was a line of students and staff outside the geography lab door waiting for me to unlock it, so that they could see for themselves the source of all the excitement.

As our river developed and we began experimenting with it, new and ingenious modifications were introduced. One of my students produced a simple scissors jack (he was adamant that he got it from his uncle who had a used-car lot) and we were able to adjust the height of the mountain at will, to cause rejuvenation or rias by raising or lowering the bedrock. We also drilled other outlets at different levels in order to control the level of the sea and in this way we were able to demonstrate isostatic recoil, raised beaches and drowned river valleys.

We added rocks and cliffs and houses and trees, all to scale and the river took on a personality of its own. Some kids put model dogs in the fields, others tried to wage war with toy soldiers, and the river was always a busy place. Another student took great pride in developing a little mechanical device that created waves for our ocean and it gave real effect to the beach areas.

It turned out to be a wonderful project even if I had more grey hairs when it was over, and one that I certainly won't forget. I know too, that the students look back on those days with delightful memories and I can assure you that they all know their rivers.

You don't have to get carried away with these 'do-it-yourself' type projects to be a great geography teacher but I must admit that, in retrospect, we had a lot of fun doing these kinds of projects and we learned a great deal. I was able to build up a selection of tools and implements that were used over and over again to good effect.

Whenever a tool was beyond the range of the school budget, we had to improvise and sometimes invent a way around it. In order to carry out cross sections and beach profiles we needed a theodolite and poles, so we put our heads together and came up with an assortment of implements that worked well to achieve our ends.

We bought a number of five-foot broom handles and carefully marked off each foot. Then we painted the alternate sections so that it would be easy to distinguish - one foot red, and one foot white. We picked up a carpenters square, a hammer (to act as a geologist's hammer), a level and a one hundred-foot measuring tape.

Cross Profile of a Beach

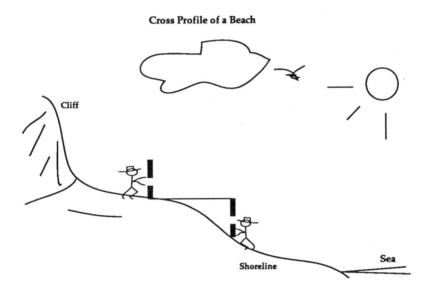

Fig. 9.2 Beach Cross Profile

And off to the beach we would go, armed (an unfortunate choice of words to describe the activity) with poles and tape, to play at geography. We drew beach cross-profiles, collected and split-open samples of indigenous and erratic rocks, studied the long-shore drift, listened to the sea while sifting through the driftwood that had washed up on the shore, lazed in the sun, and stared in peace at the sailboats and ships that drifted silently by. Geography was fun and we looked forward to the field-work.

The geography lab should have a collection of compasses to assist you when you teach direction and bearings, and you can profit greatly from a home-made weather station. Your students can conduct experiments, carry out scientific observation on a daily basis and learn to interpret signs and predict the weather with accuracy and confidence. I will dwell on weather stations in greater detail in another chapter.

The geography lab can be the headquarters and the center of activity for a number of school clubs that relate to outdoor work and scientific experiment. The school astronomy club is a perfect example and you could raise funds to buy a school telescope and carry out observations when there are eclipses and other interesting phenomena taking place. Every student should have the opportunity to look through a telescope and see the marvels of the cosmic spectacle for real, the same way that Galileo did, so many years ago.

Similarly, the mountaineering club and the kayak club could have natural tie-ins with the geography lab since they will be experiencing at first-hand and underfoot, the feel and look of the features that you discuss in your classes. Geography can become more alive for your students when they have actual first-hand practical experience on the ground. This will help your students feel at home in the lab and insure that the lab has a central place in their lives.

Geography can be learned through the soles of our feet.

Since outdoor field-work and visits are an important part of every student's geography career, you should plan your teaching schedule carefully in advance. Because of potential disruption, and collateral time-tabling impacts, it is wise to plan for minimum upheaval and try to group two or even three periods together at least once a week. In this way you can have a class that allows you two hours or even more in which to travel to a local site, carry out some work and then return to school.

Field work is a necessary and basic part of teaching geography, and you should be familiar with all existing school policies regarding these activities, or assist in drawing up new and useful policies that can facilitate such teaching techniques.

Now, its up to you. Draw up your plan and devise a schedule for establishing your own geography room. You can be the greatest advocate of your school's geography lab. Work steadily and consistently to make it the finest room in your school, a room where you are proud to teach and a room where your students are pleased to learn. Geography will become for you and your students a real subject with a meaningful learning experience and a profitable undertaking.

If, on the other hand, you are an interested outsider or a parent, please bring your ideas to the attention of your school associations, and particularly, talk to the teacher who will be imparting geography skills to your children. Your ideas and commitment do count and together you can help change our schools.

Chapter Ten:

Geography Begins at Home.

The need for a new national commitment to geographical literacy has never been greater. Despite a record level of public spending on education there remains an unacceptably high level of geographical illiteracy among our young people.

Youngsters lacking the basic skills of geographic literacy have difficulty interpreting their world as they are confronted by serious challenges in the work-place, especially in today's information age. The prognosis towards the year two-thousand is even worse. The message from industry leaders and governmental organizations is loud and clear. America is not adequately prepared for the technologically advanced society into which we have rushed, and where we now find ourselves.

We are not ready for the twenty-first century!

Industry analysts are confounded by the problem and they have perceived a need to focus on educational solutions in a last ditch effort to raise the standards of literacy, numeracy and technological competence. New and interesting alliances have sprung up nation-wide between private industry and the public school districts in an effort to help bring educational achievement into line with the specific needs of today's society. This is a good turn of events since it aligns education with the work place and will help advance American industry, but it is important that education does not sell-out to industry and the focus must always remain that of preparing our children for life, not just work. Let educationalists beware and keep the old adage clearly in their thinking.

He who pays the piper, calls the tune!

In our quest for enlightened approaches to education we must not overlook the homeplace, for it is here that the greatest effect can be achieved on our young people. After all, it is with these same young people that we must place our trust and our hopes for the future. What can you do to help? One important fact, you must realize from the outset: you can make a difference, and more specifically, you can make a difference with your own children. They do not have to grow up by trial and error, stumbling as it were, around in the dark trying to cope with a world of increasing technology and shrinking dimensions; you can see to that. You can help them become adequately prepared.

You can make a difference with your own children.

Learning is passed from one generation to the next by parents long before the child enters school. Don't ever under-estimate your influence on the education of your children, no matter what level of education you yourself have achieved. You, as parent, are your children's first and most important teacher. They begin learning at an early age when you help them use words and images to describe and interpret their world, and they continue learning throughout their entire lives. The beginnings that you provide them no matter how tentative, will do a great deal to determine how much and when they will learn.

As a parent you will have few responsibilities more important or more rewarding than helping your children learn and grow to be literate and knowledgeable so that they can take their places in their world. Stay-at-home moms have always known this, even if they have had to sacrifice careers and life-styles. And now thankfully, in this world of equal opportunity today, stay-at-home dads are finding it out also. What I am saying here is that education is not the sole preserve of the state. We cannot afford to let the opportunity slide; that of giving our children the best preparation so that they can be effective members of our global community. Geography education is a large part of that plan and is fast becoming such a dimension of increasing importance that it cannot be ignored.

I am going to go further and say that it is my purpose and the aim of this book, to help you and your children become geographically literate and sufficiently learned so that you are able to make informed decisions about your planet and your lives.

These are fine sentiments on paper, and it is easy to agree with them. But you are probably saying to yourself, "What do I know about geography? I have no formal training and I couldn't possibly teach my children something that is so totally Greek to me!"

Wrong!

Wrong!

Wrong!

You can make a difference; and the sooner you begin with your children the better they are going to be, and the greater impact you will have on their future. So make a decision to begin today. You want to do the right things and you want them to get off to a good start. But you may wonder what are the right things? Where do you begin?

To lead off, I will examine what you can do generally to help your children get the best out of their schooling and then we can focus on how that applies to geography. Incidentally, I am not making this up as I go along. The facts that I am stating here are not new, they are the tried and proven tactics adopted by parents and teachers for hundreds of years. And they work. So don't think that this is a new experiment that you and your children are taking part in, but it could be new for you.

Recently, education authorities published a series of questions so that you can help make education a number one priority in this country. Read through these questions and see how you rate with your own children. Remember that this applies to you in your own home, not in the school. And the items on this list are simple to do and a wonderful recipe for success.

1. Have you spent fifteen minutes in a conversation with your child today?

Fifteen minutes is not a lot of time. But, how many days slip by, when we are so busy rushing around trying to get this and that accomplished, that suddenly the day is done and it is time for rest. As the children grow older it becomes harder and harder to arrange to spend time with them, because they also get caught up in their own lives and the school activities and sport and games and Nintendo, and friends and so on and so forth. So try to establish a routine whereby you all have an opportunity to sit down together, as a family, in the morning for breakfast or for lunch or for dinner. Make a point of spending at least fifteen minutes on a daily basis in a conversation with your child and believe me it will be time well spent.

2. Have you read to your child today?

Reading to your young children is one of the most important things you can do to give them a deeper appreciation for life, and build a strong bond between you and them. Choose a book that will appeal to their imagination and will be a treat to share with them. It is a mistake to look on this reading activity as a chore that you must do. Get with the program and enjoy what you are doing. Read some of the great classic children stories like *Charlotte's Web*, or *A Wrinkle in Time*, or *Black Beauty*, or *Alice in Wonderland* or... There is a fortune of great stories out there and you can introduce your children to incredible new worlds and fill their imaginations with stimulating ideas and thoughts. Introduce them to folk tales from around the world; have an atlas nearby to point out the different and exotic new places. What fun!

3. Have you discussed right and wrong and religion with your child today?

This can fit in and be a good topic for the conversation piece that we spoke of, earlier. Your child should look to you for a clear message in life. What is right, what is not right and where should I go from here? As a responsible adult you can easily provide these answers to your child, but only if you communicate them to him or her. There are certain things that are not lawful and you must provide a good example in your actions as well as in your words. What message does a child get from his Mom or Dad when they tell him not to steal, and then they watch their favorite programs together looking at a stolen TV set? Or how about when Dad says never drink alcohol and then he opens another can for himself? Kids are only human. They do not respond well to mixed messages. Be straight with them. Let them know right from wrong and show them good example.

4. Have you played together today?

All work and no play makes Jack a dull boy! But you can do something about it, and before it is too late. Here's a great opportunity to bring enjoyment back into your own life. Make time for playing with your child every day. It will mean as much to you as it will to them, and you will look forward to the time you spend together. Just watch the happiness in their faces as you play together and know that you are doing right.

5. Have you gone somewhere together today?

You don't have to go far. Just put some time aside and go for a little stroll or a bike ride or a walk in the park. Keep it simple. Have your conversation with your children while you are on the walk. Play with your children while you are traveling with them.

The exercise and fresh air will do you both good. Enjoy the scenery, point out the colors of the leaves on the trees and look for beauty all around you. Don't let this opportunity, that of bringing pleasure and relaxation into your life, slip by.

6. Have you listened to your child today?

Above all, listen to what your child has to say. Remember that children are not little adults. Neither are they empty vessels that you can fill up with information, knowledge, values and ideas that replicate your views of the world. Just because you are a certain type of person and you work in a certain type of work-place, does not mean that your child will be and do the same. The Acorn must grow up to be an Oak tree and no matter how you try, you will not succeed in making it a Beech tree or an Elm. All you can do is provide them the opportunity to fulfill their dreams and wishes and be happy doing so. Listen with your ears and with your soul. Hear what is going on in their heads so that you will not be suddenly surprised one day by some awful tragic news of some stupid deed that could have easily been avoided. When they talk, listen. Listen! Listen and understand.

7. Have you hugged your child today?

You're going to love this one. At every opportunity hug your child. Hug them in the morning first thing and then every chance you get hug them all day. It is one amazing fact that people do not get too much hugging. You can't over-hug your child. Isn't that great? They will come back for more and more, time and time again so don't ever push them away. Just hug them, and then hug them again. It takes at least two to hug, but you can add more if you like. The interesting thing about it is that the huggee gets as much out of it as the huggist! I'll bet you never heard of a huggee before.

8. Have you checked your child's homework today?

School can be very traumatic for your child. So, help him, help her. Talk to them about school; in fact, make school the topic for your daily conversation now and then. Talk about what is right and what is wrong and bring up religion in school. Then listen as they tell you about their homework. I'm not saying that you should do their homework for them, for in truth that does not help in the long-run. But talk to them about it and listen to what they have to say. Take interest in their school work, and let them know that you are interested and that you want to know what is going on. They must know that if they have one of those insurmountable problems that is looming dangerously on the horizon, that you are there for them to offer a helping hand. And that is good; that's the way it should be.

9. Have you checked your child's school attendance today?

How would it be if you were to find out that one of your children was not showing up for school some days? This can only happen if you are not staying in touch with your children's daily activities and getting involved in their schooling. This is one of the most important things in your children's lives right now - schooling, and you should be doing what you can to help them out. You shouldn't have to police them but you can build up a trust and a rapport with them so that you know exactly what is going on in their lives.

10. Have you monitored and limited your child's TV watching today?

TV has changed all our lives, but we grew up with it and saw it coming. Our children think that TV was always there, and it was for them. TV is one of the most intrusive, insidious and pervasive media that exists. It sits in nearly everybody's living rooms and can run twenty-four hours a day. The problem is that on any given day, about ninety percent of what is on the TV screen right there in your living room, is injurious to the developing young mind. If you don't believe me, sit there for twelve hours and watch some of the violence, sex, language, and uncouth

things that happen every hour. You must watch the same programs that your children watch before you can stay in tune with
them and know what they are thinking and saying. If you allow
your children to watch as much TV as they like you are asking for
trouble. If you do not monitor their viewing and limit them to a
certain number of hours TV time per week, you are creating
problems that somebody will have to deal with later. Get in at the
beginning. Set up good habits for viewing TV. Make some
logical, down-to-earth rules and stick with them. Your children
will appreciate them later, if not now.

By doing these things you will find out how children -
especially your own - are growing up today. This should be
important to you as a parent. Here are a few other ideas that you
can do to really get into their world and understand what it is that
drives these kids. Ask their teacher if you can sit in the back of the
classroom, quietly, for three hours. Watch and listen, and keep
your mouth shut. Visit the juvenile judge, the hospital neo-natal
care center, and the police station. Watch six hours of cable
television in one day, including MTV. Talk to your child's teacher
and compare notes. Make an appointment with your school
principal and ask him/her the following questions.

1. Is this school drug-free?
2. Is this school violence-free?
3. What are your goals?
4. What are your academic standards?
5. Do you have a report card that measures to world-
class standards how well my child is learning English,
Math, Science, History, Geography?
6. How are the social needs of the children met so
they do not become a barrier to improving academic
achievement?

Now, don't get me wrong. All this stuff is not easy. It
requires from you, a dedicated commitment and a concerted
discipline to put the time and effort into such a program. But the
rewards will be great. This is a revolution of sorts. Not the blood
and guts type, rather the silent creeping one, but nevertheless a

revolution just the same; a revolution that will change our lives. It is a moment in time when we must interrupt what we are doing and ask questions that are of the most fundamental nature.

1. **What kind of people are we?**
2. **What is important to us?**
3. **What do we have to contribute to the place where we live?**

It is only if we each stop and make a conscious effort to make change in our educational process that change will come about. We cannot make the mistake of leaving it to the next man. We must all stand and be counted. The end result will be a country filled with able people. Every adult will be literate and will possess the knowledge and skills necessary to help us compete in a global economy. Most people are trying to do this today. Sometimes it just seems like it is such an insurmountable task, but a little at a time will take us along the correct road that leads to eventual success. I believe that people are trying to get their lives right, they are trying to bring up their children correctly, and they are trying to have better communities. But it has always been by trial and error. It would be nice to learn these kinds of skills at school.

Caring for your children, caring about their education and caring about their environment should go hand-in-hand. Education that takes into account the well-being of the planet should be an important agenda on their curriculum. You can do many different things at home to provide a solid background that will supplement their school instruction and give them a sound preparation for life.

Begin with this simple list of things that you can do today. Discuss geography related topics with your children and informally teach them about the world in which they live. Help the younger ones in their school-based activities and give them encouragement and praise. Monitor their progress at school. Become involved in their school programs, their games and activities, and especially their homework. Volunteer to take the children on walking field-trips into the country-side and also to the library, and guide them to view geography education programs on TV.

If you live in the city go visit a farm and if you live on the farm go visit a city. All this will help your child develop a keen sense of geography and a wonderful appreciation for life and nature. Children love to learn, it is as natural as eating for them. They have an insatiable appetite for inquiry and a habitual talent for exploring new and exciting places and ideas.

You can tap into this natural resource and feed it wholesome information. I'm not saying that every word that comes out of your mouth should be referenced in some way to geography, because that would defeat the purpose of the exercise and make them lose interest in the subject. Every time you talk about their world to them, you will be helping them learn new English words and increasing their vocabulary. This is good, because language frames the world that the child knows; the richer the language, the richer the child's world. In addition, you are providing a learning situation that fosters new values, helping them distinguish right from wrong, good from bad, safe from dangerous and, in there too, you can add your geographical skills. They will learn to 'look and see'.

That might seem simple and obvious but that is all you need to do. Let them look at different things and see them, not just with their eyes but also with their minds. From this simple exercise they will learn 'cause and effect' and understand the difference. They will also gain skills that will, later in life, help them deduce facts from their own clear and first-hand observations. What a great way to begin life!

During the preschool years, children develop at an extraordinary rate. Each day's experiences, however familiar to adults, can be fresh and exciting for curious preschoolers. Although your children's incessant curiosity might be aggravating, especially at the end of a long day, it provides an opportunity for you to help your children connect daily experiences with ideas and concepts. Not only are there abundant opportunities for you to help your children develop the skills and ideas that will give them the perfect preparation for schooling and for life, it can also be fun and exciting for you both.

Even if you are not familiar with the technical side of geography instruction, or even if you are not very knowledgeable at first about geographical phenomena, much of what you do at home parallels and reinforces good school instruction and your own knowledge will grow as well.

You can happily share many activities with your children that will reinforce the instruction they receive at school. The things that you are advised to do for preschoolers continue to be important for kindergartners and first graders: going places, talking about everyday occurrences and events, doing things together that build up the child's knowledge of the world and that help them make connections between words and concepts.

Things that you do together are the single most important activity you can carry out to insure your preschooler's continued progress and contentment in geography, and will be increasingly more valuable as your child moves up to school. You stretch your children's understanding of words and ideas and help them come to grips with new and complicated concepts and percepts. In addition, they learn to enjoy geography by association with happy activities and good times in your company in an atmosphere of togetherness, closeness and warmth while you encourage their natural development. Set an easy goal; find just one new thing to do with your child each day.

Yes, you can broaden your child's experience.

Talking to your preschoolers is a natural activity, but one that can make a tremendous difference to their development. There are different kinds of talk that parents "do" to their children, and I 'm not so sure that all of them are good. Some people distort sounds and words for their preschoolers and call things "baby names" as if the child could not learn the correct word. I call this 'coochy-coo talk' and it has a place in creating a certain kind of atmosphere and mood for the child. However, the same kind of mood and security can be created while you use correct sounds and words. Call a spade a spade. Your child can handle it.

Talk about the everyday experiences and events that are familiar to you both. Talk about memories and events that the child will be able to recall from the recent past. When you are going about your daily chores in the kitchen or the bedroom, talk to your child and also listen to the various babblings and attempts that will eventually mean more.

No matter where you live, your home and community can provide a rich source of experience for you to share with your child. These expeditions can be simple like exploring your home, or elaborate like going on a trip to the museum. Either way, these experiences can be rich in the history of your family and the traditions you value.

The experiences should be varied since children are especially stimulated by new messages. Providing the occasional trip outside of the daily routine can go a long way. Such a trip may be as simple as spending a day in the country if you live in the city or vice versa. Similarly, don't overlook the resources that may be available in your town like the library, the zoo, the museum, the park, the beach or the river; these are close-by and accessible, but yet provide wonderful gateways to new and exciting worlds that stimulate the imagination, far removed from your community.

The experience should be surrounded by talk. Put the experiences into words and you are building on the store of ideas, concepts and percepts that enrich their lives. Call attention to the many and everyday things that pass by. Point out objects, ask questions, and give them ample opportunity to talk and express their feelings and opinions. Help them to formulate ideas and put their concepts into words.

Be careful not to reprimand them for a mistake. Mistakes are part of learning, and you must recognize the error, accept it for what it is, and turn it into a positive and productive learning experience. Then perhaps, they won't be afraid to make mistakes in the future. Praise them a lot - an awful lot. Encourage your children to think and talk about the world around them as often as possible.

Remember, there is ample subject matter in the close proximity of your own home to, more than adequately, lay a solid foundation for your child's education.

I have said before, but I insist in re-iterating that you cannot and should not attempt to teach a subject in isolation. You may say that you are teaching geography or science or some other subject, but you need to have a working level of language in which to communicate your ideas, observations and questions. Similarly, you will need to be able to read and write, sketch and draw; organize facts and information, think and evaluate and so on and so forth. As you continue your instruction your children's vocabulary will grow rapidly, and emotional maturity will develop as they learn to formulate opinions about, and place values on, their unfolding world.

Sometimes, because of the familiarity that breeds contempt, we tend to take everyday things for granted and we fail to see the novelty and excitement that simple things can generate. Can you remember the first time you saw a beautiful waterfall, or can you recall your first skiing experience? I can get goose-bumps thinking about the first time I stood on top of Mount Kilimanjaro, in Africa, and I'm sure you have some experiences that bring a similar kind of pleasure. That kind of excitement can happen everyday for children, discovering for the first time the everyday things that we know so well. You must acknowledge this fact and encourage the creativity and excitement that the simple things in life do for them.

We do not see things the same way that they see things and this is OK. I will never be sure which is the better way to see things, our way with all the serious complications or their more simplistic 'black or white' views of life. One day my youngest boy came home from preschool and produced his latest piece of modern artwork. I must say it was a splendid splurge of colors and shapes.

I immediately praised it enthusiastically and encouraged by my reaction he informed me that it was "*Wriggly the Worm*". I thought it would be appropriate to use some scotch tape and display "Wriggly" on the door of our refrigerator, along with Kermit and a whole plethora of farm animals. I figured he was going through his animal phase. When "Wriggly" was resplendent on the refrigerator door, I stood back to admire my handiwork and my son, Ronan, reprimanded my efforts with a derisive laugh - as if to say "What do Dads know?" He howled with laughter, running to tell his mother that Dad had put *Wriggly the Worm* on the refrigerator door "upside-down". Needless to say, I didn't make that mistake again, but I laughed with him.

I didn't give up, and a few years later, I was rewarded with a different kind of learning experience. It seems like I spent the best part of my children's youth learning from them, and it was wonderfully enriching. I will always treasure the first garden that Ronan and I planted together.

It was not a large garden and I think that was good in retrospect since it coincided with his attention span and ability to focus on dirt. We prepared the ground first and then went to the store with a shopping list to get the seeds for vegetables and flowers. We decided together to plant some potatoes, tomatoes, lettuce, peas, carrots and corn, as well as some colorful flowers for Mom. We had a great day at the store talking about all the different things that cropped up.

Next day, we planted the seeds and since everything was hidden beneath the soil, we marked each row with the picture bag that held the seeds. This way the picture of the carrots on the bag at the end of one row reminded us that this was where we had planted the carrots. We got dirt in our fingers and in our ears and we were pleased when the job was done. He slept soundly that night.

Next morning, he went to see the results of his work and after a close examination, he grimly announced that nothing was growing. He seemed hesitant and uncertain at first when I told him to wait for a few days or a week, but that was back at the time when he still believed in Dad and he accepted the delay without question.

Sure enough, in a week he was able to see tiny green lines emanating from in front of the picture bags and each new day there would be more and more. What excitement! The garden was a fresh topic of conversation for us every evening after my work. He would verbalize many tales, especially relating to the onslaught of the neighbor's cats who had their eyes on our freshly-dug plot, and he could recall much of our activities as the garden grew taller and stronger. Then one by one, the buds were replaced with bright colored fruits and vegetables and we revelled in delight as the time drew near for harvest when everything would be ripe for plucking. It was good to see the fruits of our labors.

The tomatoes, the corn and the peas were the forerunners in the popularity contest, and he plucked some each day to run exuberantly into the kitchen to show Mom. As the Summer gave way to Autumn and the stalks began to wither, I mentioned one day that soon it would be time to get some potatoes. Like a new puppy he bounded with energy out to the garden and returned with a worried expression because, he announced, "There are no potatoes." He had seen us buy potatoes at the store and I knew that he could recognize one if he saw one. So I was amazed when he had concluded that there weren't any in the garden.

"Bring me the garden fork." I said, and we both retired to the garden once more. I dug under the first stalk and turned about eight fairly good sized potatoes onto the freshly broken earth. His face shone with astonishment as he turned and ran to the kitchen with a large dirty tuber grasped tightly in his hands. "Look, look Mom," he scrambled for breath, "Dad got some potatoes, and they live in the dirt with the worms."

I think I learned more than he that day. I learned that I shouldn't have assumed that he knew things that I had taken for granted. I had taken for granted that everybody knows that potatoes grow in the ground and not on stalks like ears of corn or pods of peas. That day changed my attitude about young people.

Then I began thinking: how could they know things unless we tell them and specifically show them. I wonder do kids think that milk comes from cartons and eggs from Styrofoam crates. I had a whole new perspective on life from that moment on. It occurred to me and I am amazed that despite the relative lack of planning that goes into our children's lives and education, how they still achieve a very respectable level of knowledge on their own. Then I thought if it were planned, how much better would their start in life be, and how better they would be prepared to deal with the problems and issues that come in abundance as they get older. And what parent wouldn't want this kind of start for their children?

I said before that children learn a great deal more from their parents than we think. It is important to be aware of this fact and that all your actions and deeds are by implication a major influence on your child. If you never read at home, how will your children know the value of books. If you never take the time to walk in the park with your child, how will they know to appreciate fresh air, sunshine, the rustle of the leaves, the birds, the scent of the wild flowers and all the other wonderful discoveries that are waiting around the next bend in the trail?

Especially in today's world of rush, rush, rush, it is vitally important to make time to visit the zoo or play in the park. It is vital to give them time to be young. I cannot stress enough that it is so important to plan ahead and make time to spend at these activities, because even though it may not seem like it at times, our children grow up awfully fast. And when they grow up they grow away.

Give them time to be young.

One day soon you will be visiting the zoo on your own and your young man or woman will, I hope, be accompanying their own youngsters to explore again for the first time the wonderful wild world about them. Don't miss out on one of the most enriching experiences of your life and of theirs and you will be splendidly rewarded for many years to come. When you are having fun at these activities your children grow up seeing this and learning that it is good to go to the park and play in the sun.

You set a powerful example by doing it yourself, not as a chore, but for yourself and for enjoyment and relaxation; not talking about it, but doing it, not dreaming about it, but doing it. Compare what messages a child will receive from a Mom or Dad who is always working, always too busy to play, always talking about the good times that are coming, and the messages the child will receive from you out there in the park, flying your kite or just touching the flowers. You too could be the parent who comes home from work and plonks his rear in the easy chair in front of the television, to impart sound values and judgements to your child. The answer is simple, just ask your child!

Remember that children who enter school, already knowing some basic facts about their world, generally do better than children who haven't had that opportunity. If the children are not enjoying the activity, you can do one or two things to change this; change the approach, make the activity more fun, more interesting, add excitement, curiosity, add a banana, add some ice cream; or stop, and do something else. Try another activity; it should be fun for all of you.

Keep in mind, that if children forget something that they knew last week, they aren't failures. That is just the nature of young children. Encourage them and help them grow familiar with the activity again and soon their recall powers will get stronger.

It is a good idea to start a journal and store away mementoes that relate to precious memories and happy times. You will be surprised as the years roll by how much you will value such a collection of memorabilia later on in life, both for yourself and for them. In it you could keep a scrapbook, keep a collection of photographs, cards, odds and ends, sketches and writings and anything else that you think would be worth holding onto for posterity.

The most important thing during preschool years is for parent and child to enjoy each encounter with the world they live in. This is making the best kind of preparation for geography, as well as all other learning too. Laughing about the mistakes that you and your children make is a better motivator than reprimanding your child for getting it wrong. Enjoy spending time with and informally teaching your children and be happy in the knowledge that they, in turn, will know how to enjoy spending time with their children in future years.

Ask your children questions, and then wait for the answers. Do not be a free and easy source of information for your children. I have witnessed many parents doing this - asking their child a question and then answering it themselves before the child even has a chance to cogitate over it; as if they were afraid that their children would let them down by looking dumb. This is a mistake! Avoid putting the answers in their mouths, thereby getting them into the bad habit of expecting you to answer your own questions. Many children have their parents conditioned to respond in this manner, and it fosters lazy attention and kills creativity and thinking processes. Ask the question clearly, and at the right time - when the child is focused and paying attention to you. Avoid repeating the questions and this will help sharpen their powers of concentration and attention.

Very often the parent begins with the right idea, but ends up, in their enthusiasm to be a success and because of the fear of failure, being conditioned by the child. That child is invariably a fine applied psychologist and learns easily to manipulate the parent to his/her advantage. So if you find yourself repeating and repeating in order to get a response, you need to stop, step back, and rethink your approach. It might be time to wrest back control.

State a question clearly, one time, and expect an answer.

Ask pertinent questions that require more than a simple yes or no response. Ask questions that help children to compare, contrast and classify things in the world around them. "What if" questions are important for providing opportunities for creative thinking and in stimulating the imaginative powers.

When a child asks the endless question "why?" be patient and instead of giving a quick and easy answer try saying: "I don't know. Let's look it up." This way you have shown them that books serve as a resource for answering questions and getting at truths.

When they grow up, they grow away. As children become more independent new opportunities to learn and talk about the world arise. If given a chance, children quickly learn to recognize the details they will need when they talk about an experience they did not share with you. They begin to recall and categorize information, thereby building and enlarging their capacity to remember. They will also begin to use language as a way of sharing important feelings and thoughts.

Listen to your children. Share in the excitement of their new discoveries - since it is all new to them. Even the most mundane things that have lost all spectacle of vigor for you can be wonderful sources of energy and delight for your children. They will learn quickly how highly you value their ability to speak lucidly.

The readiness is all! There is no "right" time for children to become proficient at understanding concepts and ideas. One thing you can rely on however, is that whenever a concept is proving too difficult to grasp, you can simplify the task by introducing concrete examples and tangible clues to help your child see the connections. For instance, the simple concept that water flows downhill may not be easy for your child to understand, but you can demonstrate this in actual terms with hands-on experience - admittedly, wet hands-on - but nevertheless you will create an opportunity for your child to "see" the cause and effect and understand what is going on. This is one of the fundamental building blocks for learning.

Ease the transition from home to school by giving the child ample practice at reading, writing, talking, and describing the everyday things in life around you. You can happily share many activities with your children that will reinforce the instruction they receive at school.

Trips to the library or bookstore can be pleasant outings. But be warned, these places are microcosms of the world outside and unless you prepare your visit carefully, can be fraught with anxiety. Before you take your child to the library, carefully draw up your plans; know what it is you want to accomplish and keep your goals within limits that can be attained. Otherwise your great trip could backfire and you could both end up lost and wandering haphazardly for countless hours and come home frustrated and empty-handed, with a bad taste about libraries and trips in general. Give yourself a good chance at being a success and plan your activity well in advance, then you will have less risk of ending up bewildered, and overwhelmed by the enormity of the task and the multitude of collateral material available.

To minimize this occurrence, carry out your own pre-trip planning tour to the library on your own, or with some other parents who are also trying to find their way through. Together you might hit upon some innovative ways to help your children. The other sources of help for you should be the child's teacher and the reference librarian. The reference librarians are wonderful sources of information and, in my experience, the most approachable and helpful people in the world. With their expert attention you can engage the vast resources of the information age to be your ally in the education of your child.

Become familiar with how the library is laid out and functions, and know the general location where you can find the books that you will require for your child. You don't have to know exactly where every book is located and you don't need to have a mental map of the lay-out of the library, but the more familiar you are with the reading room and the reference desk the better you will be able to ask the right questions to come up with the correct and relevant information. A well thought-out, prepared trip to the library will, in this way, be both rewarding and fun for both you and your child.

You will be in the perfect position to focus your charge on the books and information in the appropriate age group and ability. You will already know where to find the simple books with the large color pictures that relate the way the world works in an interesting and entertaining fashion. These wonderful representations of life will provide a solid basis for the development of appreciation and knowledge in geography. When you have directed them to the appropriate shelves, let them browse through the books, exercise choices and together make decisions about which books to take home.

Talk about the pictures in the books and let them express their views about what is revealed on the pages. Remember that beginning geographers, like all beginners, thrive on having someone value their emerging skills. Tell them that you are proud of them and tell them often. Show them with your actions, hugs, gifts that you are very proud of the way they are learning their new skills. You are making an immeasurable contribution to your child's future with these simple actions and words.

Don't think about it, just do it!

Make sure that there is a place in your house that is suitable for fostering learning and knowledge so that your child will be able to concentrate on the pertinent work without distractions.

Surround them with learning and they will become learned.

Have books and globes and maps and atlases in this little study corner. Provide them with ample supplies of paper, pens, markers, crayons and any other implements that are suitable for keeping them happy and busy at their work. Keep the utensils within easy reach and they will more likely get used. Keep this work-space neat and tidy and help your child develop a good attitude to organization and work habits.

Television viewing is often a concern of parents. You should place reasonable limits on TV viewing - ten hours a week is recommended for school age children. More than that has a negative effect on learning. Limiting television viewing frees up time for other activities that can be a lot more rewarding for your child. Regrettably nowadays, many parents who are overworked and worn out, view television as an easy baby-sitter where the children are quiet and they do not have to put up with their incessant questions and boundless energy.

Monitor what your child is viewing, and watch the programs with them, so that you can discuss at their level, what you have seen together. In that way, you can better help your child understand the programs. Some programs, like wildlife, natural history and science, lend themselves naturally to geography and you should be on the look-out for these and sit with your child, to share them together. This is an excellent use of the television to bring vicarious experience to your child from outside your local area, and from other places on the planet. You can foster a healthy ethos for geography by talking about these programs with your child afterwards and by asking intelligent questions about what he/she has seen.

Your most important role is in encouraging and supporting your child at school and at home. They will naturally look to you for attention - you can decide whether the attention you give them is good or bad. It is important that you stand behind the school authority by backing up teachers' requests for promptness and diligence in homework.

I am not saying that you should blindly back up every request that the teacher demands, but you must remember that the teacher is trying to create a suitable working environment for your son or daughter to excel at whatever he or she is going to excel at. The teacher's role is a very special one in the education of your family, and it can be much healthier if you and the teacher see 'eye to eye' and cooperate towards the same goals. Avoid saying negative things about your child's teacher behind the teacher's back because this does not help your child's education in the least, and it undermines the teacher's authority. Don't forget to praise the teacher when your child does well. Believe me this appreciation will be returned thrice-fold in the years to come.

On the other hand, if you feel that your child is not benefiting from a particular teacher, then it is up to you to do something about it. Go talk to the teacher and explain your feelings about what is going on. If that does not help matters, then by all means, take the case further along the chain of authority until the problem is resolved.

While you are encouraging good work and positive study habits you should not forget to applaud each success. Remember, that praise is a better motivator than punishment is a better deterrent, and children are more likely to work hard for praise, rewards and good words than they are to avoid punishment.

Have faith in your child's ability to learn, even if it seems like it's taking a long time. Some children will learn at a slower rate than others and that is OK. It is the nature of growing up. Just accept it and get on with the process of living and encouraging them.

They will learn when they are good and ready to learn and not a minute sooner.

You can help foster a good learning ethos and make it amenable for the readiness to occur. Children thrive in supportive environments and if your children get off to a slow start, it may not be because they weren't trying, but because the situation was inappropriate. Look for a situation and a place that supports your child's development.

So where does that leave you? Have you picked up any useful information from this discussion on education from the home? Here is a brief summary of the pertinent issues. These are things that other people have used and know work. But you should not limit yourself to just these skills and techniques. Use them to begin with, but remain adaptable and willing to rearrange your plan to suit your particular situation in the education of your children.

We must not overlook the home as a suitable place to mete out meaningful life encounters and education to our children. Even if you have no prior experience with either teaching or geography, you can still help and foster a keen awareness and love of the subject. There are many things that you can do to improve communication and understanding with your children so that school-life can be a happier and more rewarding experience for them. You can go further and assist them in their preparation for life by giving them a sound foundation in geography.

Begin at the beginning, with preschoolers, and progress with them as they grow up to be kindergartners and first graders. Use your natural surroundings as solid sources to enrich their learning world. Plan to take the time to allow them be young and enjoy their learning process. Let them be themselves and never judge them as failures or expect too much from them.

There is a proper method to asking questions and expecting responses - one that works and helps develop their innate natural abilities. Show them by your example and by your words. Help them by giving them concrete examples to explain difficult concepts. Enhance their learning aptitude with suitable trips that will broaden their outlook and widen their experience. Finally, set up your home to be a place of learning and the rest will naturally follow.

The geographer's home is the outdoors, and in the next chapter we will explore a little farther the educational endeavors that you can get up to with your kids, in your own back yard.

Chapter Eleven:

Your Back-yard Laboratory

Almost everybody has access to a space that they could call their back-yard. This can be a valuable addition to your learning process and could be a real asset to your children's future. The back-yard can simply be a place where you maintain a lawn or grow flowers for relaxation and pleasure; and that is a wonderful thing to do. It also could be a garden where you and your children plant vegetables, and watch them grow, or produce fresh food for home consumption.

I'm not suggesting that you try eking out a subsistence living in your back-yard, but the ability to grow your own produce is a valuable asset in today's world of mass produced goods. When was the last time you were able to say that you grew this 'carrot' yourself. You know with confidence that there were no chemicals used in its production, and no harmful sprays or pesticides that could be damaging to you, or to the environment.

I know what you are saying; "I've often thought of starting a little garden in the back..., but it takes up so much time and it's a lot of hard work." You can keep thinking about it, or you can put your money where your mouth is and do it; and I guarantee that you will spend many pleasurable and profitable hours in the pursuit of your new hobby.

But there is more to playing with gardening than simply growing fresh produce and flowers. Getting your fingers dirty in the soil is part of the deeper meaning of life - a finite and tangible connection with natural man, from whence we all have sprung. The more we progress into modernity, the greater the need to maintain our connection with the true meaning of things real and rooted in the ground. Make no mistake about it, you do your children a power of good by affording them the opportunity to experience, at first hand, the simple pleasures of experimenting with the earth and growing things for themselves.

Nature is at once the pupil and the master.

At the same time, do not overlook the garden as a great place to help your children develop new skills and learn new things. There are many other projects that you can conduct together in your back-yard that will help them excel at geography and related school work. Since education is an ongoing process and you are learning everyday, you will more than likely accumulate a great deal of expertise in areas you never thought possible. Meanwhile, you will be developing a wonderful working relationship with your sons and daughters.

A simple project that you start out with, can grow in complexity as you and your charges become familiar with the new vocabulary and the new methods. And it need not be rooted in the soil. It is your space and you can do whatever you want with it. Set up an observatory in the back-yard. Begin by conducting simple observations from your garden. Look around you and observe what is going on. Start small; pick something easy.

What kind of day is it today? How much sunshine; how much rain? Is the wind blowing; from which direction? Is this the usual direction; do you know? You don't have to be a rocket scientist to carry out these simple observations, yet there you have the basis for all scientific experiment and geographical analysis. The scientific method and geography study is based on first-hand observations, careful annotation and informed deductions as to the processes and results.

**You don't have to be a rocket scientist to make good
deductions from observations!**

Weather effects each of us in our daily struggle with life
and yet it is something that happens so much we take it for
granted. It can effect all our activities and even our moods. Are
you familiar with the sultry grouch on a grey, overcast day, or do
you recognize the bright youthfulness of people on days with
blue sky and bright sunshine. Weather also influences our dress,
our architecture and our holiday seasons. If you don't believe me
ask any Eskimo, or any Pygmy or any Australian. Sudden and
unexpected changes in the weather can leave cities paralyzed
and neighborhoods devastated, and yet we accept it as inevitable
and take it for granted. Outdoor occupations, like agriculture
based businesses, fishing and transport are particularly vulnerable to the unpredictability of the weather and people in these
professions rely heavily on their own keen observations, their
experience and especially on meteorologists.

**Red sky at night, shepherd's delight
Red sky in the morning, sailor take warning!**

You are a parent and want to give your child a better start
in education and a sound footing in geography; conduct short
and simple projects that you can all do together, that will be a
source of wonderful learning and companionship as you both
study and grow in knowledge.

You can get into the project as deep as you like; you are
not going to become an expert meteorologist overnight; and
that's OK, since there are people whose job it is, to forecast the
weather for you every day.

No matter what scientific exercise you wish to conduct,
you must abide by some rules of deliberation. Make a commitment to yourself to carry-out accurate observations for a period
of thirty days. Begin each day at the same time and carry out the
same observations; carefully annotate them in your notebook

and follow a set pattern until the thirty days are finished. It is better to not begin, unless you plan to see the project through to the end. We are too familiar with people who are easily dismissed with a cursory, "... Ah yes, she never finishes anything she starts..." It is all too easy to earn that reputation. I'm a great believer that you can do just about anything for thirty days.

Make a spreadsheet that will hold the information you observe for the time period you choose. Faithfully, fill in the blanks everyday at the same time. You don't have to be perfectly exact but it is a good idea to habituate the young person in the correct patterns of observation and note-taking.

There are a few obvious things that you must insure. Naturally, in order to observe the weather, it is necessary for you to be able to see the sky - and a good deal of the sky too. It is up there. Go to a park if it is not readily accessible outside your windows or doors. At first, you may be guessing at the answers, but little by little you will become quite the expert at recognizing strato-cumulus and distinguishing it from nimbo-stratus and so on. Each day will get more and more interesting as you accumulate more and more facts about your environment. You can get help from local newspapers and from your local library.

Methodically, fill in the blanks each day and watch your project grow. You don't have to be an outstanding geographer at first to do all this; that can come later. Look at the sky. Can you figure out how much of the sky is covered by clouds? Is half of it clouded-up; or all, or none.

The best way to measure the cloud amount is to divide the sky into eights. Use a circle and make eight divisions in it. Think of a clock. Quarter past is two-eights, half past is four-eights and so on. When the sky is completely clouded over it is referred to as "overcast". Add this word to your vocabulary list of jargon for geographic literacy. Look at the diagram and see how easy it is to represent cloud cover on simple circles depicting the sky and the amount of cloud that you estimate.

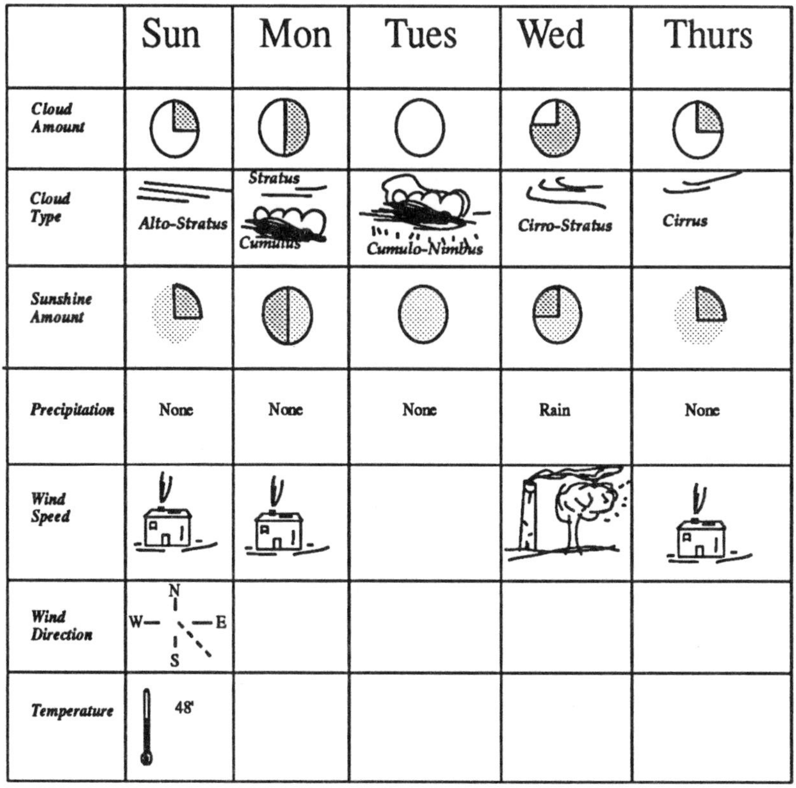

	Sun	Mon	Tues	Wed	Thurs
Cloud Amount	◕	◑	○	◕	◕
Cloud Type	Alto-Stratus	*Stratus* / Cumulus	Cumulo-Nimbus	Cirro-Stratus	Cirrus
Sunshine Amount	◲	◑	○	◕	◲
Precipitation	None	None	None	Rain	None
Wind Speed					
Wind Direction					
Temperature	48°				

Fig. 11.1 Spreadsheet of daily weather observations

You'll be surprised at how quickly you become proficient at glancing up and thinking - overcast or clear or one quarter covered. Chart this information onto your matrix and very quickly you will see patterns taking shape in the weather. This may seem stupid, but soon, cause and effect sequences become obvious. On sunny days there will be little or no cloud cover and on rainy days, ... guess what? Lots of clouds. These simple cause and effect patterns are important for young people to see and understand, and what better way to achieve this than by giving them some practical experience with observations 'in the field' as it were, in your backyard. Incidentally, can you spot anything 'funny' with the spreadsheet in the figure above?

Next, you will want to know what to call these clouds. A lot of people think these names are too difficult because they are in Latin and they remind them of medical prescriptions and doctors' signatures. But a great many of our English words originated in Latin and when you know a word's derivative it makes it all the more easy to recognize its meaning, and is a good tool to give your children as they grow up.

Here are a few easy tips that you can use to identify clouds, and the good news is that once you learn these, they will stay with you forever and they will stand up to scrutiny in all the countries of the world. You can recognize and name clouds by looking at their shape, their color and their height. At first you can cheat a little; take a stab at the name and then verify your answer by checking the local newspaper or ask your neighbors. They have already read this book and are out there right now in the wilds of their backyards, resplendent in their muddy boots, being real geographers.

Ask your kids; clouds should be white and furry or they can be black and ominous. We all know that the white ones are usual on sunny days with lovely blue skies and lots of brightness. The other picture conjures up memories of cold, biting rain and howling winds. For most people, these two categories are sufficient to deal with clouds, but with a little thought and a little observation you can know a great deal more.

The Latin for rain is - Nimbus; black clouds bring rain, so you can associate the term nimbus with dark clouds. There's a good beginning; all dark clouds have the word nimbus attached to them.

The Latin for bunched-up, or accumulation is simply - Cumulus. So now a bunched-up cloud that is dark is called Cumulo-Nimbus. That was easy. But all clouds aren't bunched-up and some are flat or stratified. The Latin term for stratified is

- Stratus. From this you can call a layered cloud that is dark - Nimbo Stratus. Cirrus is simply a thin white wispy cloud way up high. Cirrus clouds are usually associated with fair weather.

Precipitation is the word for rain, drizzle, snow, sleet or anything that falls from a cloud. It can be observed and it can be measured; and when you put these measurements with other data you can accurately describe the weather in a given location. For instance, if a place gets three inches of rain a year you would imagine that this was an arid region, whereas if a place received thirty three inches per year, that it was extremely wet.

Weather Station

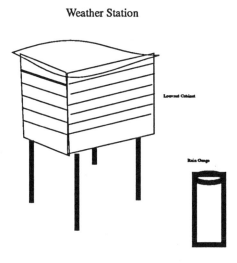

Fig. 11.2 Home-made Weather Station

I'm not suggesting that you begin keeping accurate recordings of rainfall or anything, but it is a good idea to give your children hands-on experience at collecting data and charting results. The information becomes their own, and is not just something that was read in a text book. This way, it is never boring. There are a host of other reasons also, for giving your children this type of research practice and all are excellent in promoting the kind of learning habits and skills that will be of paramount benefit for them in later life.

Place an empty container in your garden and wait till it rains. Measure the amount of water in it on a daily basis, using a ruler to measure in units of inches. When you suggest this exercise to your child you will no doubt be greeted with a good many legitimate reasons why this wouldn't work. "The rain might evaporate in the sunshine between showers," or "someone might pour water in or spill some out," or "A dog might come along and...." And of course they are right; and these are good questions and sound observations.

There are a thousand good reasons why your experiment might not work, so you have to place your container carefully so that it has a chance to collect the rainfall. Isn't it the same for all aspects of life? Aren't there always a thousand good reasons for not doing something, but if we always acquiesced to our fears, when would we ever get anything done? That's why I say; just, do it!

In real life, meteorologists had to contend with these same questions and they devised a rain gauge that accurately reflects rainfall amounts. It is designed to measure rainfall carefully, since its sharp edge can cut a drop of water, and its peculiar shape prevents any droplets from splashing out. Usually, weather stations are secured by fences that keep out strays like dogs or children or aspiring geographers who might interfere with the readings. It is not too difficult to get accurate readings and you'll be surprised how quickly your store of first-hand observations will grow and soon you will be able to produce informative insights into your own weather patterns.

You could build your own weather station in your backyard. All you need is a rain gauge and a thermometer to start with and later on as you learn more about maximum and minimum readings and relative humidity you could add-on and grow with the process.

Can you explain the difference between weather and climate? When you have your first week's readings charted neatly and colorfully for all to see, you are well on your way to telling the difference. Weather refers to the atmospheric condition in a certain place at a given time. Climate, on the other hand, refers to the average state of the atmosphere over a period of about thirty years. Begin today, by studying the weather and one day you could be the expert on climate in your local region. Look at the state of the atmosphere where you live right now. In order to determine its condition and give a description of your present weather, you might need to examine the temperature of the air, tell the barometric pressure, check for rainfall, note the wind direction and speed, and finally plot the cloud cover and sunshine amount.

This exercise is a wonderful way to play at geography, at home with your kids in the back-yard. But if you are a teacher and you wish to inculcate geographic skills and the scientific method in your students this is a perfect opportunity for you to begin a creative program with your 3rd/4th graders. By the time they are seniors and about to enter the real world outside school, they will have experienced practical utility science and geography.

The daily observations and the resulting charts will become a major achievement not only for them but also for the school. It will be something that they in turn can pass on to their children, from generation to generation. Every geography laboratory should have a weather station that the students build and maintain together, and every child should have an opportunity to see and understand weather and figure out how climate happens. It is a great way to teach 'cause and effect' and a perfect way to pass on very useful skills.

It was a dark and stormy night... and it was lucky that you were able to tell the force of the wind. We all know that wind speed and direction is measured with wind vanes and anemometers, but who can afford to have this kind of expensive scientific equipment in the backyard; and who needs it anyway! When was the last time a hurricane or a tornado swept past your house; or swept your house past?

The good news is that you don't need to buy any expensive equipment to measure the speed of the wind and you can very easily calculate direction with a little careful thought and planning. You can judge the speed of the wind with the aid of simple memory maps that allow you classify different forces. The accompanying diagram (Fig. 11.1) depicts everyday scenes that should be familiar to each of us. When there is no wind the smoke rises straight up from the chimney stack and the tree is still. When there is a gentle breeze this is reflected in the smoke angle, as it leaves the chimney; the tree sways slightly. But when there is a strong force, the smoke is horizontal as it leaves the chimney and the tree is struggling to remain upright. These are simple illustrations that help us classify the force of the wind and they serve to give us a mental image of a phenomenon that we cannot see.

It is fun to draw and color these diagrams and then it is rewarding to apply your knowledge to everyday occurrences. Once you establish where north, south, east and west is, in your neighborhood it will be easy to name the wind. A wind that blows from the west is called a west wind. Try it for yourself. What would you call a wind that blows south east? Of course, ... that's easy - *"a north-westerly"*!

After a while, when you and your child are out for a walk you may hear a comment like, "It's pretty windy today. Look at the way those trees are bent over and the smoke from that chimney is blowing horizontal. It seems like it is blowing from the north, and that explains why it is so cold today." Of course you would respond with; "Yea, and I think it's going to rain, I see that the sky is overcast and there are a lot of cumulo-nimbus clouds up there." Hey, now you're talking! Already the vocabulary has increased and new concepts are in vogue.

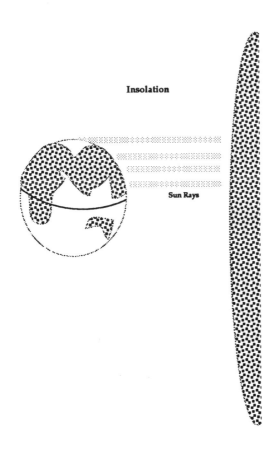

Fig. 11.3 How the Sun provides Insolation for us

What causes the weather? The answer can be summed up simply in one word - the sun - but then you could spend the rest of your days coming to grips with the complex interactions between that far-off star and our planet. If the earth was flat, things would be easier to comprehend, but it is not, and consequently they are not. Heat from the sun is referred to as solar radiation and it is the prime cause in the temperature variations that result in movement of air masses. Recently I was on a flight over the Midwest; the pilot had to make a detour to avoid an 'air mass' which we could see from our in-flight vantage point. I guess I had read about atmospheric conditions for years and I thought I understood the concepts, but when I saw, at first hand, the enormity of this stormy body of air, I suddenly realized the colossal forces that were involved in our atmosphere.

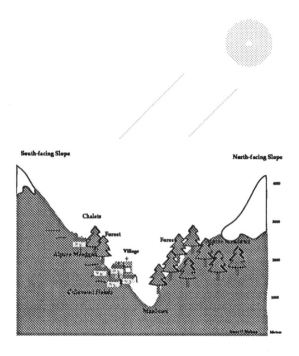

Fig. 11.4 Insolation in an Alpine Valley

And yet these forces are governed by some pretty simple scientific principles. The easiest one for me is that 'hot air rises'. I know that because I have observed it at home in my house, from the ventilation system and from watching the kettle boil. The sun heats the earth and the earth in turn heats the air above it. This is called *insolation*, not to be confused with the other word bearing a similar spelling and pronunciation, but meaning something wholly different. Insolation is the amount of sunshine you receive; insulation is protection from something. I suppose you could be insulated from insolation, if you hid your house in a forest or something like that.

Certain areas are hotter than others, because they receive more insolation than others. This occurs for a number of reasons and brings about considerable effect to the climate in each region. The intrinsic shape of the earth causes areas directly beneath the sun to receive more intense insolation than areas that are remote, like the higher latitudes towards the poles.

Sea and Land Breezes

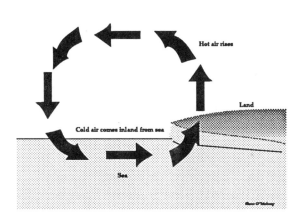

Fig. 11.5 Land/Sea breezes

The aspect of a place causes a local variation in the amount of insolation that takes place. Many people take this fact into account when they are purchasing property and setting up a home. Aspect is the direction towards which a slope faces. For instance if you are facing directly into the sun your chances for getting more insolation are better than if you were on the other side of the hill facing away form the source. This is reflected also in land use activities especially in alpine or mountain valleys where the slope facing the sun will be more apt to harbor settlement and intense cultivation than the other side, which is usually just wooded. Notice in the accompanying diagram (Fig. 11.4) that the snow line is lower on the slope that is not facing the sun. Cause and effect - the real truth!

As you know, the earth's surface is not uniform, that is, it consists of both land and water surfaces. This complicates things further since areas of land and bodies of water absorb and radiate heat in a different manner and in differing amounts. Here is a wonderful opportunity for you to do a little experiment with your class or with your child at home. Set up a situation where you can measure the rates at which equal amounts of land and water take-in and give-off heat. The results will link directly into the causes for much of the unrest in our atmosphere and our daily weather patterns. Water, because of its liquid nature, reflects much of the solar radiation and changes temperature slowly. This means that it heats up very slowly, but at the same time it cools down very slowly also. Land, since it is solid on the other hand, absorbs heat rapidly and gains temperature. At the same time, it gives off that heat readily and cools rapidly. This causes land and sea breezes, something you are familiar with, if you like to sail after work in the evening.

During the day, the land heats up fastest. Hot air rises and cold air from over the sea comes in to replace it. This is a sea breeze and serves to cool the coastal areas on hot days with a refreshing scented flow. The opposite occurs in the evening and on into the night when the land cools down rapidly. Then air over the sea is warmer; it rises and will be replaced by colder, denser land air. This simple sketch (Fig. 11.5) is a microcosm of the monsoon world where intense equatorial heating occurs and vast pressure differences exist because of the imbalance of the temperature of the air masses.

The Monsoon is simply land/sea breezes on a grand scale.

There are some common occurrences that effect our daily existence, but are rarely understood and are easily explained by insolation and related issues. Why is it colder on cloudless nights? And what causes a build up of smog? Since the air receives heat from the ground by radiation, it makes sense that if there is no cloud cover more radiation can take place, with a resulting heat loss into space. Clouds tend to bounce the radiation back down to earth and a double effect is taking place. Even in summertime and in deserts clear nights can be quite chilly since the protective insulating blanket of clouds is not present. Smog is caused by the simultaneous occurrence of a number of unrelated factors, and can have serious impact on our health and life-styles. First, the land gives-off its heat continuously for a number of nights resulting in a layer of cold air by the ground. Warmer air and radiation fog rise to the top of the cold air where it forms a layer of low stratus cloud. Lack of wind and sometimes surrounding high mountain ranges compound the situation, so that there is no relief. Meanwhile, a mixture of deadly smoke, industrial and exhaust fumes build down from the top, unable to escape the natural trap.

All this geography from a simple beginning in your back-yard! The key to understanding the entire planetary interaction is available to you, right there at home. You just need to ask the right questions, make the correct observations, apply the solutions you have at your disposal and the wealth of world knowledge is yours for the taking.

This is why I say; do not overlook your own back-yard as an opportunity for yourself and for your family to learn critical facts and theories about humankind on its planetary journey through life. Not only will it be fun and creative for you all, but it will stimulate sharp thinking and promote positive conversation in your midst.

Begin today. Give it a try and watch it grow. In the next chapter, you will discover another useful tool for improving your geographical literacy and enhancing your knowledge of the earth.

Chapter Twelve:

Maps

No matter what you do or where you go maps are an invaluable source of information and direction for you. I sometimes wonder how people survive without maps; but then, I suspect that even people who do not realize it use maps everyday. Whenever we go someplace we have to use a mental map to help us navigate there; going to work, or school or play you, too, must use a mental map to get places, give directions or explain your surroundings.

I'm sure that there isn't one of us who has not had a bad experience either receiving or giving directions in the past. Can you recall the person who, try as he may, succeeded in complicating the instructions and getting you completely lost, frustrated and confused when all you asked was for simple directions to go somewhere or find someplace. It is at times like this that you suddenly wish you had a map. "Turn left at the second light, or was it the third... no, right at the top of the hill, ... kitty-corner from McDonalds or Ernst...." Is there anything worse than this kind of nightmare when you are in a hurry to get somewhere and you know that you are just around the corner.

A map is a graphic portrayal of a part of the earth. Maps have lines, words, symbols and colors that show the distribution and arrangement of features upon the earth's surface. Each feature is drawn in a reduced size so it can be shown on paper; one inch on the map may show one hundred miles on the ground.

Almost everyone uses a map at one time or another. They help us travel from place to place and to understand the world around us. They help us plan vacation trips and follow news events in all parts of the world.

There are two types of map; a *general reference* map and a *thematic* map. The *general reference* map shows general information such as continents, countries, rivers, cities and other features. The most familiar *general reference* map is the automobile road map.

Thematic maps emphasize some particular feature, such as rainfall, the distribution of people, or particular kinds of crops. There are as many kinds of *thematic* maps as there are features whose locations need to be shown for some reason. People use thematic maps to learn how different parts of the world vary in many ways.

Maps on which different colors indicate various countries are called *political* maps. The relative size and arrangement of countries are easier to remember when shown by colors.

Maps that emphasize the characteristics of the earth's surface are called *physical* maps. Sometimes color indicates elevation above sea level. Shading is often used to suggest mountains and hills as they would appear in real life. Colors may also be used to show differences in rainfall or temperature. Colors or symbols may show the distribution of vegetation or where various languages are spoken.

The amount of information you learn from a map depends upon your ability to read it. Maps are drawn to scale in order to be accurate. The scale must be shown so that you can use the distances on the map to figure out the real distance on the ground. A large-scale map covers a small region in great detail so that you can see roads and make out small details like bridges and houses. Small-scale maps, on the other hand, cover large areas and consequently, cannot show much detail. Each map is useful to exploit its advantages and you should be aware of the scale that would suit your own purpose.

The scale is expressed in three different ways, and some good maps will have all three displayed right there where you can clearly see them. There is a very good reason for displaying the scale in three formats and you already have a preference for a particular one because this is how you can best understand it. Other people naturally may not agree with your way but they also can choose their favorite method and everybody is happy.

The first is a graphic scale, a straight line on which distances have been marked. Each mark represents a certain number of miles on the ground. The scale is also expressed in words and figures with so many units on the map equaling so many units on the ground. This might appear on the map as one inch equals fifteen statute miles; one inch on the map equals fifteen miles on the surface of the earth.

The scale can also be expressed as a representative fraction (RF) for those of us who prefer to see things in mathematical formats. This is the most common method of expressing scale since it transcends linguistic barriers and is an international mathematical language. A scale may be written as 1:62,500 or 1/62,500.

The advantage of this method is that the scale is expressed as a fraction regardless of what measurement system is used. On your map, for instance, one inch may equal 62,500 inches while on a German version of the map one centimeter might equal 62,500 centimeters. Both measurements are of course accurate.

Using symbols makes it easy and possible to display a large amount of information on a single map. The important thing to remember is when you use symbols you need to include a legend or key, that explains the meaning of each particular symbol, so that everybody else knows what's going on.

Some symbols represent cultural features of the landscape such as highways, railroads, farms, dams and cities. Others represent natural features such as mountains, lakes and plains. The symbols may be lines, dots, circles squares, triangles, words, letters, colors or combinations of these. To be most effective the symbol often looks like or suggests the feature that it represents. For example, some maps may use a tree symbol to display a forest or an orchard. This is sometimes further refined by using a coniferous tree to depict evergreen forests indigenous to a particular region and a deciduous tree to show trees that lose their leaves in the fall and one of each would show forests containing a mixture of both coniferous and deciduous trees.

General Reference maps show cities and other features; Aeronautical Charts provide information for pilots, showing airports, compass directions, and radio station frequencies. U.S. Geological Survey Maps show the exact location of features on the ground in great detail and with precision accuracy, since surveyors rely on these maps for specific facts that have to do with their measurements and work.

Maps have splendid practical application for everyday use. One of the most important skills for living that you can impart to your children is the ability to glance at a map and know what is going on. Where are you? Where are you going? What is the best way to get there and how long will it take? We take many things for granted in our daily grind but the ability to be able to navigate our way through the maze of modern-day living is a skill that saves us countless hours and numerous headaches. It is a survival skill at its best.

Map-reading is a useful international tool for daily navigation.

How can you instill this great ability in your children? Is there something special you can do to help them grow up aware and knowledgeable in the art of map-reading? Of course there is. The best thing you can do for them is show them by example - that is, know how to use maps yourself, and take every opportunity to get them to help you find out how they work. Talk about maps to your children. This should be no surprise, but when was the last time you made reference to a scale or legend or to a key? You must focus on these issues and make them part of your family's daily routine. You cannot expect the children to bring up the topic themselves. Use maps deliberately in front of them; go out of your way to make them notice. Get them involved and let them explore the legend and work out the scale. Work together on direction and distance and help them to orient the map so that it accurately reflects their concrete footings. On a daily basis it doesn't take much to engender a positive elementary interest in mapwork about the house and you can take heart in knowing that your children are acquiring useful international skills that will serve them for life.

But beyond that there are a great number of interesting things that you can do to foster a healthy interest in maps. Begin at home. Start small and let them get used to the methods and the concepts before you tackle any bigger projects. Let your child make a map of your living room or a bedroom. Title it *"My First Map of My Bedroom"* and color it, so that it is something that they can be proud of, and would want to display. All you need is a measuring tape, some blank paper and a pencil, and you should choose a suitable scale so that the real measurements on the ground will fit clearly and manageably on the paper. If the scale is too small the measurements will not fit on your paper and if the scale is too large the resulting map will be so tiny that it will not be suitable for viewing. Read this again carefully, since the majority of people seem to be confused by scales. It is quite simple; the smaller the scale the larger (meaning more detailed) the map, and the larger the scale the more country you can fit onto the page, but the less detail.

Small scale, more detail
Large scale, less detail

I suggest a scale of one inch to two feet, and see how that works out. You might have to experiment a little to hit upon the correct scale that suitably represents your particular room and effectively displays the doors and the windows and the furniture and the other things that your child might want to add in. After all it is her room and her map, so why not show the things that she wants to show? There must be a space for the doll's house, the cat or the dog. This is a living geography and it helps if it relates specifically to the child's world, so be creative and encourage active participation in the project. Make it fun.

A Plan of My Room

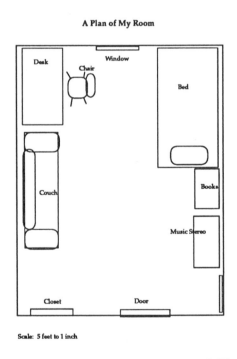

Scale: 5 feet to 1 inch

Fig. 12.1 Plan of My Room

When you have successfully created your first map of your room it is time to make a map of your house and garden. This takes a little more measuring and careful note-taking so that you don't have to do it over and over again. It is a help to first draw a free-hand rough sketch and then add the measurements to this so that it makes sense. In this way your child is gaining practice at creating conceptual maps and mental images and then verifying the facts in the field, or the back garden, whatever the case may be.

Careful planning will bring its rewards later.

Again, as before, spend some time in coming up with the ideal scale to represent your drawing on the paper size that you have to work with. It is a good idea to make the map almost as large as the paper that you have to work on and remember to leave room at the top for a good title, and at the bottom for the legend and the scale.

Take the time to draw the finished product accurately and neatly. It must look the part, when you put in so much effort, and try not to be satisfied with a half-hearted attempt that does neither of you any justice. Stay with the project until it is done. Get into the habit of seeing a project through to the end and collecting the just reward for the results. If the project was put together carefully and given the time necessary for completion the results will be ample to justify a good response from all who see it, and will be a source of joy and pride for the successful map maker or burgeoning cartographer.

Stay with the project until it is finished and collect your just reward. You've earned it!

When the map is complete, use it. First get the children to orientate the map so that the house and the map make perfect sense. This is the first thing that you must always do when you open a map - orientate it so that it relates to the ground you are trying to read. If you have a compass, mark north on the map, and make a mental note to familiarize yourself with where north is, in relation to your front door. When you step out of your house what direction are you facing? This may seem trivial, but as a geographer you should know these simple things as a matter of form.

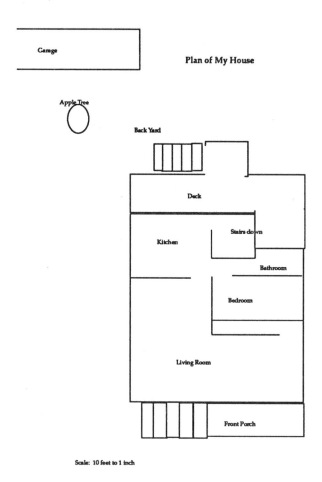

Plan of My House

Garage

Apple Tree

Back Yard

Deck

Stairs down

Kitchen

Bathroom

Bedroom

Living Room

Front Porch

Scale: 10 feet to 1 inch

Shane O'Mahony

Fig. 12.2 Plan of My House

Let your children get some practice at making mental maps and using them. Ask them to describe an everyday journey that you undertake first in words and then ask them to write it down. Next, get them to make a rough sketch of the journey, say from home to school and finally, go out and physically verify it. This exercise will help them to come to grips with the workings of maps and scales and direction. But what about you? Could you make an accurate depiction of your journey to work or to the store? Go ahead, try it.

You can be creative also about what to map with regard to the other projects which you have begun as a result of reading this book. Why not make a map of the weather station that you just set up in your back-yard! You can have a lot of fun and provide a very useful skill for your children's education at the same time. We haven't even mentioned yet about the possibility of job preparation in the future. As more and more awareness is being fomented about geography, and as mapping techniques are becoming highly technical and computer generated, there will be a keen demand for highly skilled, ambitious geographers in the years ahead.

> Wanted immediately: Geographer to manage GIS
> operation. Experience with ArcInfo systems and
> Database management.
> Must have degree in Geography, Town Planning or
> related field. $4,000 per month!

Wouldn't it be pleasing to be able to point your children towards a livelihood that is not only rewarding and fulfilling, but also one that they truly enjoy doing. You can foster that love of geography from an early age by doing just a little on a daily basis. Begin today!

If you are in a classroom situation, one of your first priorities must be to give your students practice at map orientation and some mental dexterity with mapping concepts. This you can easily achieve on the first day back after vacation when the students are becoming familiar with their new classroom and new surroundings. Mental images and maps are being created on that day at any rate. You can capitalize on this by introducing them to a mapping game that will help them develop their intellectual facets and help you individualize the children. You will also find out pretty quickly, who has difficulty with the spatial concepts and directional issues with which you are working.

The Geography Lab

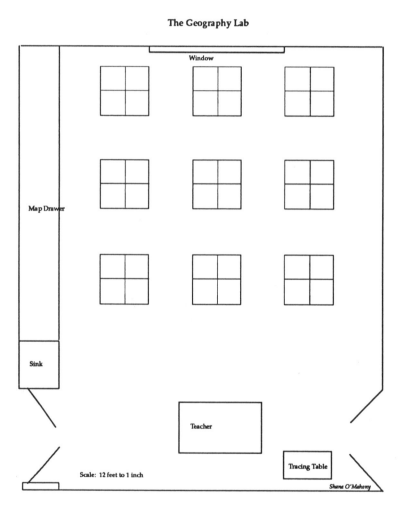

Fig. 12.3 Plan of My Geography Lab

All you need to do to play this mapping game is prepare a plan of the geography room (your present class) and include each pupil's desk or chair. Clearly mark the teacher's desk and leave the rest of the map blank. Now pass a copy of the plan to each student and ask them to write their name on the position where they are seated. In order to accomplish this simple task each student will need to orientate the chart so that the plan and the room show the teacher's desk in the same place and then they need to figure out where they are seated in relation to the marked desk. You will learn many amazing facts about your class from a little exercise like this one, and you will find that it is time well spent.

Once you have everybody oriented, motivated and suitably enthused about their spatial whereabouts, you can ask each student to draw a plan of his/her desk using a suitable scale. This is sufficiently vague to invite different examples, and you will be in a good position at the end to demonstrate the advantages of one scale, that shows the desk in a suitable size, over another, that may show the desk so large that it won't fit on the page or too small and of little value visually.

Insist that each individual place a clear, legible title on the map, preferably on top, and likewise an accurate scale at the bottom. In fact, it is a good idea to get them to construct all three scales so that they fully understand each one and can expound on the advantages of one over another. If there is a need make sure that they place a legend on the page so that any symbols that were used will be adequately explained to the new map-reader.

Finally, ask each student to sign and date the work and instruct them to leave a space for the teachers signature at the end. This will get them into the habit of taking responsibility for their own work, and recognizing that it is their creation and something to be proud of in front of their peers. This helps build up their self-esteem and pride in their work, as well as laying the foundation for cooperation and trust between you and your students for the future. It gives you a real opportunity to praise their work individually as you swing by each desk to sign their latest masterpiece.

The next mapping project could be a plan of your class room or geography lab. This is an ideal project for a double class period and for small group work or functional teams. One team could decide on a suitable scale, while another team is measuring the dimensions of the walls, doors and windows, and still another is bringing it all together by drawing the finished plan on the overhead projector transparency, for all to see. This is a wonderful way to get the class to work together in a positive environment as a team, and to create a useful end-product at the same time.

Once they have figured out how the process works and how very useful it is to be able to represent a large space on a single sheet of paper, they will be ready to take on home-work projects that are related to this work and will enhance their map making abilities as well as their mental-map capabilities. You could get them to map their everyday journey to school, or create a plan of their bedroom at home, or let them have a free choice - a plan of something that they'd like themselves.

Finally, you could undertake a larger project, like mapping your school buildings and play-fields. And you could chart the local shopping center or many other local areas in your school's hinterland. Rest assured that there are plenty of available sites waiting for your avid young cartographers to chart and scale, with legendary success. Get out there today and plan your first expedition. Just do it! This is where geography flourishes at its best. I am going to describe a team effort that we conducted in one of our younger classes, when I was willing to experiment with methods and materials.

It turned out to be a very effective mapping project that captured the imagination of my students and inspired them with zeal for creativity. It was also useful for developing teamwork in carrying out a major scientific study in our geography lab. Here's how it all came about and you can try this today.

The idea is to create a detailed map of your school and environs. To accomplish this, you need to construct a small scale that will facilitate representing the environs on a composite paper screen that is as large as one wall of your classroom. This will of course place the school (with the geography lab) at the center and each of the students homes should be in the surrounding hinterland, as well as their respective routes to and from school. This point should not be overlooked. It is important to conduct the experiment in the students' relevant surroundings. You are employing a sound educational principle here, one that facilitates learning, by ushering your students, with ease, from the familiar to the new, from the particular to the general, and from the concrete to the abstract. By placing the student at the center (along with his house, his neighborhood and his school) you are increasing the actuation of ownership, and you are enhancing motivation that stems from personal involvement and interest. In short, you are setting yourself up for success and your students with a meaningful learning experience.

Learning is easiest from the particular to the general.

This is a big undertaking and needs careful consideration at the outset for time, equipment and commitment. There are those who would admonish you and say - "Better not to begin rather than not finish" - and I think that, where school projects are concerned, it is indeed better to not bring the subject up unless you are about to make a guaranteed commitment to its completion. So if you know that you will not have the time or the dedication to bring this project to fruition with a suitable conclusion, wait until next term or let somebody else so it. For this project, everybody in the class needs to be involved and teamwork must be a considerable proportion of the enterprise.

As you know, a small scale, detailed map will be a pretty large depiction - it could well be the length and breadth of one wall of your classroom. Obviously, as a consequence, there will be a number of physical constraints that you, as a team, will have to consider, in order to explore feasible solutions to bring about the culmination of the product. Make a list of the relevant issues and formulate a plan when you sit down in your discussion groups.

Problem solving is a realistic, practical outcome of this type of education and can only work if everybody is allowed contribute their opinion, in an atmosphere of free discussion. It is important to act as facilitator at these meetings and be careful not to produce the answer in prepackaged format, just because you already have experienced the problems and the results. Allow them the luxury and the opportunity to make their own mistakes, so that they can, in turn, learn from them and in doing so, get a good foundation for the real world outside the school. Let the children do the exploratory work themselves and arrive at the conclusions by trial and error if need be. This is the only way that they will acquire ownership of the project and get behind it as the driving force. After all, this is how real life projects get founded, funded and accomplished.

What equipment do you need to create this map? Where do you begin? Who should do what? These are some of the questions that need to be answered before you lead your cohorts off into the sunset with theodolites and compass in-tow.

This is a classroom project and you shouldn't have to require any special equipment beyond the normal everyday classroom issue. This must include a wall, some large sheets of blank white paper pasted together with scotch tape, an overhead projector and of course a local area map. You will undoubtedly find use for a lot of standard office type appliances during the course of the project, but they are readily available in your geography lab anyway.

To begin, get someone to trace the local map onto an overhead projector (acetate) transparency. Be mindful that the school must be at the center of this map so that you can display all the students homes and routes to school from all directions. At the same time another group can be taping the sheets of drawing paper together to create the large screen. This needs to be pinned onto the wall so that you can next focus the overhead projector image on it.

When the transparency is ready, place it on the overhead projector and focus it on the 'paper screen'. Move the projector farther away or nearer so that you can squeeze all the edges onto the screen as needed. Now it is time to let the artists go to work to trace all the route-ways, blocks, and so on around the school. You will probably learn early-on, that it is worth establishing a boundary line (or crop marks), so that if anybody accidently moves the projector or the screen you will be able to re-establish the original position and continue the tracing without impacting the end-result.

It is important to allow each individual the opportunity to plot in his/her own house and garden so that they get to feel excited about their 'own' map. This enhances the feelings of ownership and pride in the project and carries them through the unpleasant more laborious tasks. Refer to the planning committee for input about the appropriate legend, title, scale and color scheme as the map gets closer and closer to completion.

When the job is finished, stand back and admire it. Take the time to praise the workmanship and marvel at the ingenuity, the teamwork and the commitment to conceive, manage and implement such a useful undertaking. Make notes of any rec-ommendations that could have improved the techniques or made the job easier for future classes.

This is a good time to have a camera with you at school. Of course you could have been taking discreet photographs from the beginning, depicting varying stages of the project as it neared completion, but be certain to take a team picture at the end when your map is created. Collect this moment and save it for yourself. Know that you have accomplished a fine and worthwhile project and one that will benefit your students in their lives ahead.

Save this class picture for yourself, but also, save it for the school yearbook and for posterity. I'm sure that the children will treasure that photograph in the years to come when they will recall the energy and fun that they shared with you, in their geography lab.

Direction is a very important part of map reading and there is only one way to truly learn a sense of direction and that is to try it. Go out into the unknown world and learn to read the signs that are all around us and that tell us where we are pointed and what lies ahead. We need to develop these direction sensors so that we can navigate our way around the uncharted escapades of our lives.

A compass is a very useful tool to help us get orientated and every pupil should at some point in their lives be given the opportunity to gain a degree of familiarity with how they work and what their uses are. I will quickly describe a few simple compass simulations and games that you can quite easily use in your home or garden to demonstrate the rudiments of compass points. Later, you can graduate to greater tasks if you are so inclined and even devise your own games and past-times.

First, and based on the premise that - what I do, I understand ... let us make our own homemade compass on paper. It may seem obvious to you that north is exactly opposite to south, and east is opposite west, but young people need to get hands-on experience at figuring this out and seeing how it all comes together. Let them draw a compass. Begin by making a large circle; place North at the top and locate East, South and West at 90 degree intervals marked in large bold letters - N E S and W. Remind them that the full circle contains 360 degrees and that the coordinates of the compass points are calibrated in divisions of this.

Next, establish the intermediate points between each of the major positions. North-east between north and east, north-west between north and west, and so on around the circle. Finally, they must insert another grade of measurement between each minor angle; north-north-east between north and north-east, east-south-east between east and south-east, and so on around the circle until your compass is completed. This may seem simple, but believe me, it takes a concentrated effort to create an accurate compass rose and have it visually accommodating also. However, you will find that the time will be well spent and a greater understanding of the workings of angles and directions will result, for all concerned.

The Compass

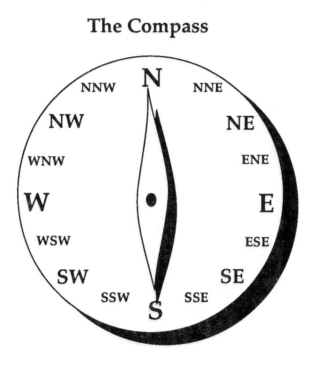

Shane O'Mahony

Fig. 12.4 The Compass

Place a real compass in your hand so that the needle is pointing north. This is, of course, magnetic north and you know that you have to make a little calculation to fix true north. The calculation is only a matter of addition or subtraction, depending on where you are located on the surface of the earth, and is not a big deal by any means. However, it is important to know all this and to demonstrate how a compass helps one to locate his/her real position on the ground. Incidentally, you do know, don't you, why there is true north and magnetic north? This is a good time to explain it to your children. The compass needle points to magnetic north, which does not coincide with the geographic north pole, and this has a real bearing on navigation.

Since the needle in the compass is attracted to metal objects, you need to hold your compass away from any such objects when you are reading directions, else you will not get accurate readings. I recall how once, on a forced night-time climbing party, a group of us were making our way over rough terrain in mountainous landscape, and our fearless leader was boldly forging ahead confident of his compass readings.

Naturally, we were hopelessly lost and it was only when daybreak eventually replaced the cold, dark night that we finally discovered the extent of our bleak ordeal and realized our predicament. With the morning it also dawned upon us, that through-out the night, our illustrious leader was holding his compass next to his flashlight while he was navigating farther and farther away from civilization, away from warm food, dry clothes, sleep and home. That was an expensive compass lesson to learn and one that I will not forget for a while. You too can discover your bearings in this crazy fashion, or you can take your children into the backyard or the park and partake in a pleasant game that simulates the early explorer and the expedition leader at work with his navigation compass.

Place a coin in the grass at your feet; it works better if the student uses his own coin. Point your compass in any direction and walk that bearing for a number of paces - say sixty paces at a bearing of twenty-four degrees, making sure to always point the needle towards the north. Stand. Add one hundred-and-twenty to your bearing and walk that line for sixty more paces. Stand, and do it again. Add one hundred-and-twenty to your bearing and walk that line for sixty further paces.

This should bring you right back to where you started, since three times one-hundred-and-twenty equals three-hundred-and-sixty degrees, and that is one circle. Pick up your coin. It should be right at your foot. If it is, you did a good job at walking the bearing, and if it isn't.... it could be a costly lesson!

You can change the numbers and repeat the same game a few times to get more practice at walking accurate bearings and to find more coins. After a while you will probably figure out that it is best to point the compass in your desired direction and transfer the sighting to the horizon and fix a point there. It is easier to walk toward that point than to walk while looking at the compass needle. Walking straight lines is not as easy as it may sound. Try it blind-folded first and then try it using a compass and talk about your findings. Compass games are great fun and practice brings rapid improvements to your form and many times more pleasure to the adventure. Soon you will be ready for orienteering or other navigational sports.

Treasure hunting can be a fun game also and a great adventure as well. They can be conducted in schools or in the home, and there are games that suit adults as well as games for younger children. It is up to you, since you choose who plays and you devise the game.

In order to bury a treasure you need to draw a map and have a title, a scale, a legend that shows the symbols you used to depict rivers, bridges, roads and so on. These are the basic elements of all maps and you can copy them straight out of your atlas or any map that you find lying around at home. Make sure you insert a north line that identifies where true north is and then you will be able to give directions and issue instructions for your treasure hunting accomplices to go find the site of the buried treasure.

Here's a general idea of a simple treasure hunt sampling:

> **Begin at Harry's house in coordinate A 6.**
> **Travel north along Rte 73A for 3.4 miles until you**
> **reach a swamp.**
> **Walk NE through the swamp until you reach a river.**
> **Name the river.**
> **Follow the river until its junction with the tributary**
> **Wabash. How many miles is this?**
> **Go to the first upstream bridge over the Wabash.**
> **From here proceed North West for 2.8 Miles to a town**
> **called Alice. South-south-west of the town is a mixed**
> **woodland area and in the far corner there is a hollow**
> **where the treasure is buried. What is the name of this**
> **hollow?**

With a little ingenuity you can add a great number of features that are by nature, artistic, geographical, scientific, or entertaining, to the treasure hunt that will stimulate the imagination and foster more learning and questions.

Map work is exciting, fun, and useful.

As you and your children become familiar with, and gain confidence in navigation and compass work, you might even explore some entertaining hobbies that feature your new-found skills. These include orienteering, mountaineering, astro-navigation, sailing and a host of other related activities. You can get hooked very easily and look forward eagerly to your spare-time, when you can participate in new adventures and exciting discoveries. And then again, you could carefully choose a new career that involves some, or all of these great skills, where you spend your time doing something you really enjoy and make a decent living at it, to boot.

Chapter Thirteen:

Rivers

One of my earliest school-time memories has to do with stories relating to the source of the river Nile in Africa. I remember I was fascinated by accounts that relayed the wonderful adventures of the early explorers, Livingstone, Speke and Burton, as they journeyed into the unknown continent. I guess youngsters like this kind of stuff. I was particularly enthralled by the tale about Livingstone's death from malaria; his body was brought back to Britain, but his heart was cut out by the local tribesmen and buried under a tree on the shores of Lake Tanganyika. What was the name of the British explorer who was sent by the New York Herald, to find Livingstone in the year 1871? There in a town called Ujiji, nestled under the equatorial sun on the shores of that lovely lake, the now-famous words were uttered. Stanley, was it not?

Doctor Livingstone, I presume?

On further reflection, it is not just Africa that attracted me. I was also deeply moved by Mark Twain's river stories and especially by Huckleberry Finn. I could disappear into my reveries for hours with his characters being wafted along on the great Mississippi. Rivers do that for me. One of my favorite hobbies now is white water kayaking, but I also delight in floating lazily with the current from river bend to murmuring river eddy on the White Salmon or the Rogue.

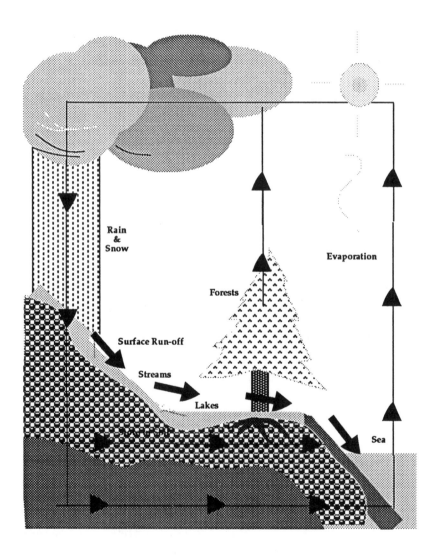

Fig. 13.1 Hydrological Cycle

Where do rivers begin? I asked my twelve year old this question recently and I was shocked at his attempts to respond in any way that would convince me that he had been to a school lately. I explored the matter to greater depth, but only confirmed my earlier amazement. It appears that our children are acquiring some pretty strange ideas and concepts concerning rivers and water and lakes and geography and God... but that's another day's work.

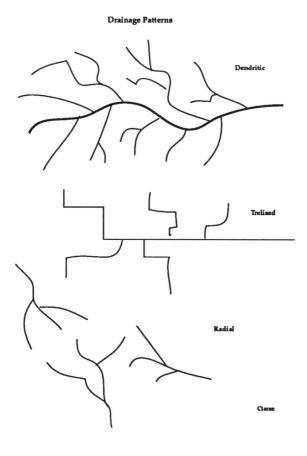

Fig. 13.2 Drainage Patterns

When it rains, where does the water go? What is runoff? What causes rivers to be the way they are? There are so many interesting questions and concepts that are pertinent to fluvial activity that it should be easy to motivate young children to find out more about river action.

Rivers start small. Each little stream flows downhill, making its circuitous way towards the sea. This little stream is called a tributary. Rivers flow in valleys and the valley's shape can tell you quite a great deal about the stage that the river has reached. A small river that has begun up in the mountains usually has a V-shaped valley with a narrow floor and sloping sides. This tributary joins up with other tributaries and together they form a larger tributary or a river that has grown in volume and speed.

Rivers form patterns and systems. A pretty normal pattern resembles a leaf from a tree, with one central vein and a lot of smaller off-shoots merging from either side. The Greek word for tree is 'dendron' and we refer to this pattern of drainage as 'dendritic'. Look carefully at a leaf; hold it up to the light and you will see that there is a central artery running the length of the leaf and it is joined from each side by lesser veins all in the same direction. In a similar way, a river forms the central artery and the tributaries merge with the main stream in a downhill direction and flows on to the sea or a lake. Look at a physical map, that is, a map which shows rivers and you will be able to pick out the dendritic drainage patterns. There are other patterns also depending on the topography and the underlying geological structures of the landscape. The most common are trellised and radial, see the diagram (Fig. 13.2) for examples.

Upper Course

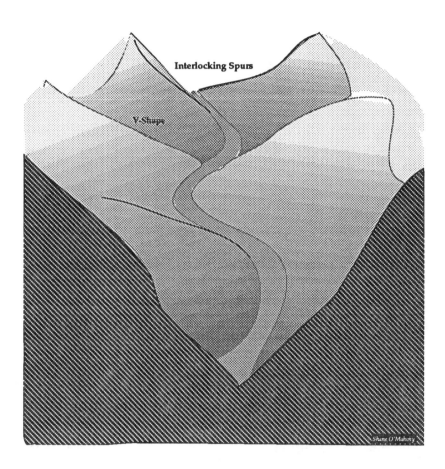

Fig. 13.3 The Upper Course of the River

Rivers begin way up on the mountains and flow downhill in a valley until they reach the sea. They can rise from a spring high up on the side of a mountain, or flow from the meltwater of a huge glacier. Sometimes they begin in corries that are entrenched near the top of the mountain. Rivers may first pass through a flat body of water, called a lake, on their way down to the sea.

Rivers have three clearly distinguishable stages as they follow the downward path to the sea. A river can be in its upper course, middle course, or lower course and each stage has different signs and clues associated with it so that you can quickly and easily discern where it is in its journey.

The upper course is in the mountains where the water rushes and tumbles over a stony stream-bed - a sound that is refreshing and soothing to all of us. At this stage the gradient is steep, the water is turbulent, and there are waterfalls and rapids and many, many small tributaries that join to form the main branch. Very often the channel will be dissected and split by flows and islets resulting in a braided stream. The river flows in a steep V-shaped valley and when you look up-river you see interlocking spurs of landscape as foothills blend into the river valley. The interlocking spurs are a clear indication that the valley was formed by a river and that the river is still in its upper course. The same valley would be U-shaped if ice had carved it and the interlocking spurs would be truncated.

The upper course is characterized by erosion as the river is down-cutting on its way towards the sea. There are three ways that a river enlarges its area due to erosion. Primarily, by vertical erosion it deepens its channel. Secondly, through lateral erosion, the banks are widened, and finally through headword erosion, the river 'creeps' uphill into the mountains and lengthens its channel in an upstream direction. The sides and floor of the bed are eroded and worn away by the hydraulic action of the water. Hydraulic action is where the force of the water wears away the sides, banks and bed of the river. Differential erosion results in many spectacular features and makes our landscape wonderfully picturesque. Rivers erode their courses and they also help sculpt the general landscape, since they tend to migrate over-and-back across the valley, as time goes by.

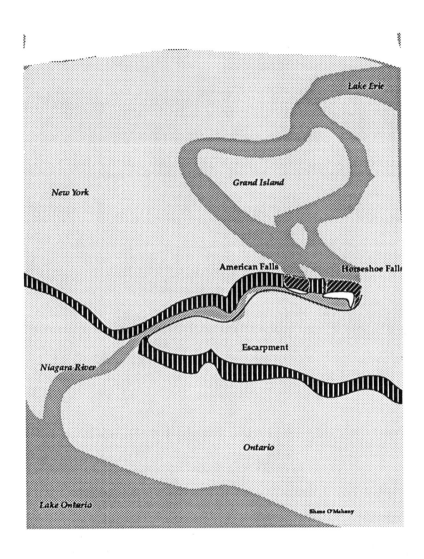

Fig. 13.4 Niagara Falls

Pebbles and boulders are picked up, bounced and trundled along in the flow; these scratch and abrade the sides and bed of the stream. This bouncing and trundling along in the flow is called *saltation*. They also scratch and abrade each other and this is why you always pick up smooth, polished pebbles from river beds and along the beach. This process of erosion, where rocks wear each other to a smooth finish, is called *attrition*.

Long Profile of a River

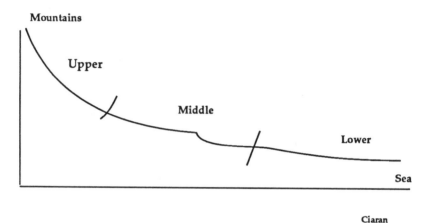

Ciaran

Fig. 13.5 Long Profile of a River

Erosion is a most powerful tool used by rivers over the ages, as they wear away the surrounding countryside and abrade rocks to make those strikingly beautiful formations that we can cherish today. Some of the most wonderful scenes of earth sculpturing are all around us; places that are readily accessible where we can view, first hand, the tremendous power of water at its work. I'm talking about the Grand Canyon, Niagara Falls, Yosemite and many, many other natural phenomena that are incredibly spectacular to behold. There is no legitimate excuse for not seeing these places. Every school-going child in America should be allowed the wonderful opportunity to verify the breadth and beauty of this country first hand. No ifs, buts, or maybes ... just do it!

Go visit the mountains and the rivers. Enjoy nature.

Make a pact with yourself today, to visit the wilderness parks and see for yourself the beauty and vigor of our country-side. Let's take a close-up view of a wonderfully spectacular feature of the upper course of most rivers - the waterfall.

Niagara Falls is a perfect example of the hydraulic action and differential erosive capabilities of a river. It is equally popular for its tourism attractions nowadays, but I prefer to envisage the original river, devoid of the trappings of modernity that have exploited its endemic natural beauty. The Niagara river is not a very long river, but it does have a remarkable incline. It is only about three miles long but drops a total of six-hundred feet in that distance. I have mentioned differential erosion a number of times when I discussed hydraulic action. All this means is that water erodes soft rock faster than hard rock. You may well ask, what is soft rock, since you could observe that all rock is hard - after all, rock is rock. Would you prefer to be hit on the head by a soft rock as opposed to a hard rock? Yes, you're quite right, but some rocks are more resistant to erosion than others. These are called hard rocks and the rocks that are more easily eroded are termed soft rocks.

The Niagara hinterland is made up of layers of sedimen-tary limestones, sandstones and shales. The top layer happens to be a layer of hard limestone and is more resistant to erosion than the softer layers beneath. For this reason, the top layer extends beyond the lower layers in many places. Behind the American Falls lies a hollow, called the *Cave of the Winds* which was etched into the cliff by the aggressive torrents cascading over the rim under the extended shelf of hard limestone. The water tumbles and plummets into the gorge two hundred feet below.

Niagara Falls is a combination of some ice age features and some river features superimposed on the original landscape. Huge lobes of ice gouged and scraped the depressions that today comprise the Great Lakes. Two of these lakes, Erie and Ontario, were separated by a steep two-hundred foot escarpment. An escarpment is a geographic name for a slope that drops-off on one side more abruptly than on the other. It is a wonderful word and you should make note of it to add to your growing geographical vocabulary.

During the melting stage Lake Erie overflowed and the Niagara river carried the excess water over the escarpment, and began to excavate the gorge that we see today. The waterfall was situated at a town called Lewiston at that time, but because of continuous abrasion the gorge has been eroded back seven miles to its present location. The river is typically braided and has resulted in two separate and distinct falls, one for each nation - The American and the Canadian falls. There is more water flowing over the Horseshoe Falls on the Canadian side than over the American Falls because of the configuration of the channel around Goat Island, and this means that the Horseshoe falls is being eaten away at about three inches per year, a rate three times faster than the American Falls. There is an awesome sense of energy and uncontained power as one stands on the bank beside the falls, literally inches from the edge as it plummets over the rim into the gorge below.

The undercutting and collapse, caused by the continuous pounding of the water, highlight the erosive power of the river and give us a glimpse of the majesty and splendor of nature. You can experience the same invigorating sensation on the ledge at Murchison Falls, in Africa, or Angel Falls in South America; in fact, any place where the upper portions of rivers display their magnificent treasures. After the upper course comes the middle course and again here, the rivers display different characteristics and accomplish another kind of work.

When the river passes from the upper course to the middle course there is a change in gradient. The river is less steep as it emerges from the mountains and moves out across the plain, on its way to the sea. Sometimes, if the change in gradient is abrupt enough, a feature called an alluvial fan is formed at the juncture of the upper and middle courses. The river has to slow down and in so doing deposits much of its load, which in turn is sorted and spread into the characteristic fan formation. The most notable thing about the river in its middle course is its increased volume since all the tributaries that merged with it earlier add to its size and capacity and give the river more power. At this stage too, the valley is wider and deeper and the gradient is less steep. The river does a great deal of work here also in the form of transporting the material that has been eroded already.

Middle Course of the River

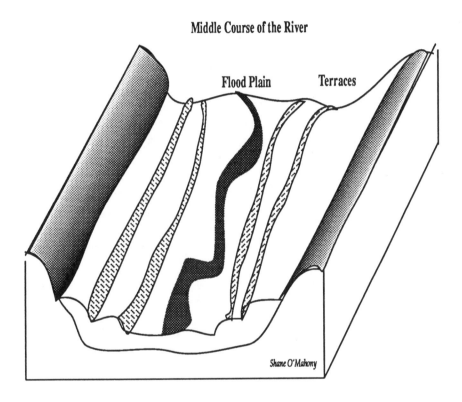

Fig. 13.6 River in its Middle Course

It erodes the bed and banks as before, by abrasion and by hydraulic action and it also causes the loose rocks and pebbles to be worn down by attrition, but the main force is used for transporting the eroded material on down towards the sea. It does this by trundling along the boulders and debris that are on the bottom; but it also dissolves mud and sand and carries it along in solution form, by a process called *suspension*. You can tell if there is a lot of dissolved silt and mud in a river because it will have a dirty brown color; this is usual after heavy rains, when the river level is high and there is a danger of flooding.

Sometimes a river may look clean but in reality there may be a good deal of silt suspended or dissolved in it. You can do a simple experiment to find out how much material is in the water. Collect a jam jar of water from the river and inspect it. It may look clean enough, but allow the jar to sit for an hour and check it again. Notice the sediment that filters to the bottom and settles. If you collect a specimen from a very muddy, raging flood you will get a good sample of dissolved and gritty mud. When the jar has rested for a day or more you will notice that different layers are quite visible and you will see that you can distinguish between the heavier and lighter materials simply by looking through the glass.

Here in the middle course of the river the channel is wider due to lateral erosion, and a flood plain is formed on the floor of the valley. A flood plain is what we call the low-lying hinterland of the river that is likely to be inundated when the river bursts its banks during the rainy season, or when the snow thaws higher up the valley in the snowfield or on the glacier. Accumulations of alluvium, that is sediments and mud deposit from the river, tend to build-up over successive flooding seasons and will enrich the soil in the valley in the long run. The middle section of the river is also marked by developing meanders that give it a characteristic river look and feel.

The river doesn't flow evenly since there is a drag from the bed and sides and turbulences are set up. Erosion is encountered and deposition occurs at different places and the result is that the river migrates across the valley, widening and deepening it as it proceeds towards its final goal - the peneplain.

As the river flows rapidly down towards the sea it reaches its lower course and there are clearly defined characteristics associated with this stage also. The lower course is usually the stage before it reaches the sea, where the river slows down or stops. When it slows down it begins to drop its load, sometimes called detritus, and the associated work is the major activity of the river in this lower course. This is known as deposition.

Lower Course of the River

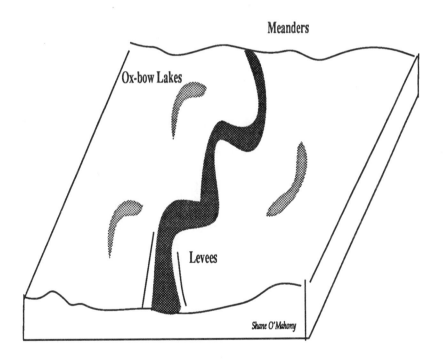

Fig. 13.7 The Lower Course of the River

The migration that originates in the middle course is more widespread and easier to spot at this stage. The river continues to meander and classic examples of slip-off slopes on one bank with cliff under-cutting and collapsing on the opposite bank result in huge meander loops and abandoned cut-offs or ox-bow lakes. Slip-off slopes make great beaches. It is good to get an aerial view from a plane, of a river in its lower course to see the abandoned channels and intricate patchwork of waterways that develop when the river is winding its way to the sea.

Tributaries fed the river at the beginning and distribu-taries process the water through the many mud-flats and silt deposits at this other end. From your vantage point in the air, you will be able to pick out the bird's foot delta where the river dumps its load at the confluence of tide and flow. At times you will also see that the river at this stage is very susceptible to flooding and careful observation will enlighten you as to why.

The river continues to erode to sea level and when it reaches that level it will drop all its load. If that happens before the river is done reaching the sea, it will build-up its bed and continue down to the sea. Flooding will occur sporadically when rainfall has been heavy up-stream in the catchment area. Every time the river floods, it widens out and slows down, thereby depositing a thin layer of silt over the land. After many such occurrences the river will have built itself a channel and banks that are higher than the surrounding countryside. These are called *levees* and they are to be seen in the lower course of river valleys. A good example is in the lower Mississippi valley, where there is a very unstable floodwater situation, since a breach in a levee can cause thousands of acres of flooding with resultant damage to life and property. "Bye, Bye Ms. American Pie!"

The solution to the Mississippi river as to other rivers with levees is to be found up stream in the catchment area. Runoff must be controlled and tributaries need to be curtailed so that the floods down below are not catastrophic. Back in the sixties a major scheme was invoked to rectify the problems that excessive flooding, mosquito infested swamps and poor agricultural techniques were causing. The *Tennessee Valley Authority* was implemented over a period of twenty years to control the flow of water in the catchment areas of the major tributaries like the Ohio and the Tennessee. Agricultural technology was improved by introducing contour ploughing and arresting soil erosion; hydro-electric stations were brought on-line to increase the power supply to the national grid. The project was a geographic solution to a natural phenomenon and greatly helped man conserve the landscape and at the same time harness the potential offered by nature.

Sometimes when land movements outside the scope of river action coincide with the erosion and deposition cycles of rivers a remarkable and wondrous spectacle is produced. The Grand Canyon was formed when isostatic recoil caused the land to rise at the same time that the river was down-cutting into the channel, which in some cases was meandering. The result is an entrenched drainage system with spectacular deepened meanders, and now you can visit these places and view, with pleasure and understanding, the magnificent sights.

In many instances, I draw for examples extraordinary visions of grandeur and splendor to illustrate a meander or a waterfall, but I want to reassure you I do that only because they are there and I have been lucky enough to have visited them. However, I have spent many pleasurable hours at the beach playing with tiny streamlets that gurgle to the ocean and there I have been able to see and understand the wondrous workings of our planet. The same principles of science govern the little stream as the mighty river and you can recreate the waterfalls, the rapids, the braided stream, the delta and any other features you care to on the beach.

Neither do you have to fly to Arizona or to South America to find them. Remember, geography is all about you and you are as good as you want to be. Don't be afraid to find the beauty in your midst.

Rivers are the epitome of geographical features. Nowhere else will you get to observe the entirety of the geographical spectacle in all its breathtaking splendor. Land-use activities and people's utilization of nature's resources are intricately tied to river topography. Cast your mind to any river you know and think about the activities that people engage in, along the banks of this river. Take any one of the tributaries of the Mississippi and follow its course, from where it rises up in the mountains to where it enters the sea. Let your mind envisage this long path. Let us pick the Tennessee, for instance. Contrast the people and activities back in a picturesque Tennessee valley with the daily lives of people in New Orleans. What a fascinating and diverse example, and yet they are all part of the same river system and all share in the same fluvial journey.

Look across at the Missouri river now, and note that the same diversity occurs there also, as economic and cultural factors interact with the physical topography to create the pattern we are familiar with. Apply the same inquisitive eyes to any river system in any part of the world and you will be able to discern similar patterns of behavior and economic activities that are, in fact, universal. Just think! How many of the world's major cities are situated on rivers? What happens when a river rises in one country and passes through several more before it enters the sea, like the Rhine or the Euphrates? This is where you could learn valuable lessons in economic geography, as you discover how political borders are not necessarily contiguous with physical boundaries. One nation's exploitation of the river could have far-reaching implications for neighbors farther down-stream.

For me, it is a magical equation, to be able to verify again and again in all parts of the globe, where ever I rest a while, that the same sounds emanate from a rushing, gurgling stream in Tibet, as they do in up-state New York, or in Outer Mongolia, or in the mountains of Wyoming. I'm glad we study rivers, since the more we know about them the more likely we are to preserve them. I hope that you get the opportunity to share this remarkable pleasure with your family.

Chapter Fourteen:

Ice

Recently, I read in a scientific magazine where a couple of enterprising adventurers were harnessing Alaskan icebergs and "bottling ice" to sell, at exorbitant prices, to the Japanese consumer. They were adamant that the ice was three times colder than normal *refrigerator* ice and lasted three times longer. I guess that's why they could charge ten times as much!

At any rate, it got me thinking about ice and I brought the subject up over dinner with my teenage sons, only to learn to my chagrin, that ice comes in cubes or as slush and is very necessary for certain drinks. I should have known better, but ... one lives in hope. My youngest boy had a "cool" idea - he said that we should put an outboard engine on an iceberg and drive it from the north pole to the Sahara desert. There it could be melted down and we could use it to irrigate the soil, with which to grow grain that could be used to feed the starving millions, who are experiencing famine conditions. Go for it, I said, but only to convince myself that I was part of their generation! I love these kind of practical everyday solutions that boggle the mind.

It seems to me that our young children know precious little about the effects of the recent ice age and cannot equate the topography today with major movements of ice in the past. And yet all our lives are touched by what has gone before and we shouldn't have to live under the shadow of such phenomena in darkness. This is especially true since the concepts are pretty straight forward and easy to assimilate with what we already know in relation to rivers and mountains.

Ice left widespread remnants and imprints on our landscape because of the powerful ability of glaciers and ice sheets to erode, transport and deposit material. Glacial features are the cause of much of the land-use patterns that have developed in modern agricultural practices and very often the work of ice offers reasons for the location of mines, cities, waterways and routeways. Nevertheless there is much apathy toward the study of glaciation and there is a blissful ignorance regarding the processes that engrave our topography.

What about yourself? Are you one of these people who is not very knowledgeable about glaciation? Never fear, you are in good company! We should learn about ice and glaciation at school, if for no other reason but to be able to interpret our surroundings when we travel and understand why things are the way they are. Much of our landscape has been sculptured by the action of glaciers and ice sheets and in many places today the ice is still very active and we should know what physical activities are involved and what cause and effect processes are taking place. I'm a great believer in knowing as much as possible about my surroundings. It is especially important for us to know everything about our planet, so that we can survive to the best of our ability on it and preserve it in its true state for our children. How else will they be able to enjoy the delights and the resources that we have had and still have now.

There were at least three ice ages that resulted from climatic deteriorations over long periods, and lasted each for several million years. The last ice age began about two million years ago and lasted until about ten thousand years ago. Doesn't it seem ridiculous to be speaking in tens of millions of years so flippantly, especially since modern man is only a couple of thousand years on the planet? It makes us pale into oblivion when we think of the enormity of time involved in just one ice age. But it is good for us to see things in perspective like this. During each ice age there were periods of advance and recession which were separated by long interglacial periods when the climate was very similar to what it is today.

Now there's food for thought! Are we at the beginning of another ice age? There are those who would point to the recent deterioration in climatic conditions and prove that we are, and of course there are those who would disagree and say that we are headed for the opposite - a *green house* effect, resulting from global warming and damage to the ozone layer. And really, I don't think that you or I will be around to find out who is correct. Will we freeze-up, or will we burn up? Or does it matter? We will know for sure in a couple of million years.

Today, ice covers a large part of our planet - roughly ten percent - and plays a distinct role in all our lives, especially in the lives of those who are directly impacted by its daily frigid reminders. But we are all, in some ways affected by the colder regions of the planet since Polar influence over weather patterns and climatic conditions result in direct impact on people in nearly all parts of the globe. Try swimming in the waters off the Washington coast even in the middle of summer, and when you have recovered from the shock you'll know what I mean. You will probably never forget the cold Alaskan current that flows south from the Bering Sea bringing chilling sub-arctic waters to the shoreline. That same cold current has another effect on the inhabitants of California further south. Cold marine air comes into contact with warmer air from the valley and is one of the prime reasons why San Francisco has a dismal Summer, one that prompted the now famous remark by Mark Twain or words to that effect.

The worst Winter I endured, was the Summer I spent in San Francisco.

We know that parts of the planet experience torrid conditions since they are directly under the sun's rays in tropical climes. At the other extreme, there are remote areas where the sun rarely shines and they experience arctic conditions.

Other places seem to be anomalies. They are in direct sunlight right on the equator, and yet experience arctic conditions. Some mountains on the equator are this way. In the beginning these places like Kilimanjaro, perplexed early travellers and explorers who weren't sure how to explain ice in the middle of Africa on the equator. Kilimanjaro was first explored by a German traveler, named Gillman, in the eighteen hundreds. His report, that of seeing a glacier on the equator, was laughed at by his peers in Europe and he was told to get out of the sun since it was thought that too much would have adverse effect on the brain's normal functioning.

Be that as it may, we now know that temperature decreases with altitude, dropping approximately one degree centigrade with every three hundred feet of incline, and that explains why mountain peaks can be situated on the equator, in Africa or in the Andes, and still have glaciers on them.

Two kinds of ice conditions exist in our world: continental ice sheets and rivers of ice called glaciers. While they are basically the same, since they are each made of ice, they differ in extent and size. The continental ice sheets can cover entire countries and are typically located around the poles. Glaciers are localized rivers of ice that flow from the mountain peaks down through valleys to the sea.

At different historic times ice sheets of various sizes and masses proliferated over the greater part of the earth. Today we can see the tracks and scars that these huge excavators left behind and with a little training we can see the patterns that explain whence the ice came, and where it went.

If you are standing at the north pole the only way you can look is south. So it follows that ice from the north pole flows south. Think about this for a minute! The ice takes the path of least resistance where it can, and where it can't, it simply moves the obstacle out of its way like a monstrous bulldozer - pushed on from behind by an enormous force of weight.

How does ice leave its mark on the landscape and how does that effect you? Let's take a closer look at some of the physical characteristics of ice and some of the processes that take place in glacial conditions. Much of the landscape of the modern-day North American continent is the result of sculpturing and excavation by massive ice sheets that spread from the polar regions. Lake Michigan, for example, was formed by a giant lobe of the arctic ice sheet that lay over the land ten thousand years ago. Constant advance and retreat of that ice lobe bulldozed and gouged the hollow, and when the ice melted it left in its wake a huge lake. The center of the ice was situated around the area we now know as Hudson Bay, and here ice piled deeper than ten-thousand feet in thickness. The pressure of this great weight caused the continental ice sheet to flow westward and southward so that it covered most of the land mass of North America, down to about the present valleys of the Missouri and the Ohio rivers.

Ice is heavy, and it moves downhill under its own weight. In fact, it is so heavy that very few things can stop its movement and anything that gets in its way will be crushed, moved, bruised and scratched so that it ends up very much changed from what it started out to be. An ice sheet that begins at the polar regions builds up layer over compacting layer until it finally has to move downhill from its source under its weight. Ice moves because of the force of its own weight, helped by gravity and naturally because of the inherent viscosity of its make-up. After all, when we think of ice we immediately think of slipping and sliding and skating and frozen lakes and ... More and more snow falls in the arctic regions and the bottom tiers become compacted into neve or hard ice. Many names that are associated with glaciation today come from the French, you'll be glad to know. This is so, because much of the early study of glaciation was conducted by Rene Descartes, Elysee Reclus and others in the French Alps during the last century.

I taught school in the outback of Australia, one year, and I will always remember the interest the children at the local school demonstrated when I brought up the subject of glaciation and snow. They had vague ideas that snow fell from the sky - but since their skies were rarely anything but bright azure-blue I could see how they could wonder. One astute kid asked if the snow would "bong you in the head, when it fell from the sky." I'm sure that you have been "bonged" in the head by large hail-stones from time to time but you associate snow with soft flakes.

However, next time you get the opportunity, go into your backyard and pick up a handful of freshly fallen snow. Look at it carefully and see the flaky structure and notice how light it is and how much air it seems to hold. Now, squeeze the handful of snow until it is fully compacted and compressed, thereby expelling all the air. You will immediately notice that its color has changed and its size and texture have also changed. It is quite hard now and indeed if you throw this snowball at somebody it could sting quite a bit.

Suppose you continue to scoop up more and more snow and compress it all into a larger snowball, you will see that it becomes heavier and changes to ice. Children learn early that compacted snow makes a better snowball than loose snow, and their parents are always wary when they figure out how to make a real humdinger that could injure somebody's eye or do some other damage. What is the difference between loose freshly fallen snow and hard compacted snow? In a word - air! One is snow surrounded by a lot of air and the other has no air. It has turned into ice.

Ice has other qualities also that are simple to digest, but when you envision what can happen at a grander scale it is easier for you to understand the processes at work and we begin to comprehend why the landscape turned out to be like this. For instance, did you know that when water freezes it expands and its volume increases in size by about one eleventh or roughly by ten per cent. So what, you might think!

Here's something else to think about. If ice is really water in another state, how come it floats on water? What do you understand by the expression, "the tip of the ice-berg"? The answers to these questions are simple but by thinking about ice in this manner, you will begin to comprehend how our landscape ended up being chiseled and sculptured by the 'mighty bulldozer' for millions of years.

Ice Berg

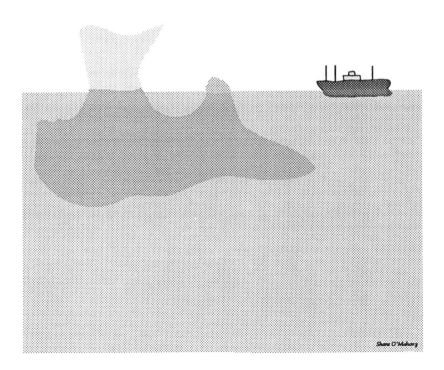

Fig. 14.1 Iceberg

Expansion makes ice lighter than water and this is why it floats. Ice may float on water, but approximately ninety per cent remains underwater and dangerously, out of view, since out of sight - out of mind.

Water freezes, expands and that is why you need to put anti-freeze in your car's radiator each winter and it is also why you might need to wrap your water pipes in insulating material, lest temperatures fall below freezing point. If your car's radiator or your home's water pipes froze the water would increase by ten per cent, causing the radiator or the pipe to burst. Not only would these be very expensive to replace or repair but it could happen again the same day if you didn't take precautions to forestall the freezing.

Notice that when snow or ice is on the ground there is always a wet patch under it and around it. The water seeps into the ground and sometimes when it is cold enough the ice freezes solidly to the ground so that it could only be removed with great force or difficulty. Transfer this principle to the ice sheet that sits over northern Alaska and Greenland and other places within the Arctic Circle.

The ground under the ice and snow is wet and when it is cold enough (which can be just about all the time) the ice is frozen solidly to the ground underneath it. We mentioned earlier that the ice sheet moves out from the polar regions, forced to do so by its own weight and by gravity. Since the ground beneath it is frozen solidly to it the ice rips it up and carries large chunks along with it. In this way, huge boulders and rocks become embedded in the base of the ice and they in turn act as the sharp, cutting teeth of a giant rasp that scrapes, scours and scratches the landscape as the whole agglomeration moves over it.

These scratches are called *striations* and people who are trained to read these marks can tell many thousands of years later, where the ice was, whence it came, and in what direction it was going. Much of our landscape today has been formed and chiseled by the movement of ice, and we can understand the processes that took place and know why they occurred by looking at striations and other marks.

For instance, since all this snow is falling at the poles and the ice sheet is growing thicker and thicker, a number of things must occur. First, all the water that would normally fill the lakes and feed the rivers is now locked-up in ice, sitting atop the continents. Remember this water should be flowing into the rivers and down to the sea. Yet the clouds that bring the snow pick up their water vapor and moisture from the oceans and when this depletion continues to take place over a prolonged period of time it is natural that the sea-level should fall dramatically. Scientists have estimated that the ice age beaches were at a level three-hundred feet below the present shorelines.

As the continental ice sheets move away from their central gathering grounds, the one from the north and another from the south, it seems obvious that they should all coalesce at the equator. But wait! Ice melts at temperatures greater than freezing point and this prevents the ice sheets from moving where ever they like on the face of the earth. It also explains why the ice sheets stopped moving forward at certain latitudes. The snow was still falling and accumulating back at the source and the ice sheets were advancing all the while. So where was the ice going?

The ice sheets were melting. They began to melt at the front where the ground temperature rose above freezing-point, which technically meant that the continental ice sheets were moving but not advancing. This mammoth melt created a great many rivers and lakes that originated and lasted only for the duration of the ice age. Then, when the ice age was finished they disappeared, leaving in their wakes the dried up valleys and forms of former waterways.

Sometimes the ice trapped huge lakes between natural barriers and an artificial dam caused by the wall of ice. In many cases the lakes lasted for thousands of years and thick layers of silt were deposited at the bottoms of these bodies of water. This happened across Saskatchewan and Manitoba in Canada and prepared the way for rich fertile agricultural land today.

In other places, the artificial dam that retained the water body, suddenly gave way under the enormity of the pressure and released monumental cascades of water down through the local valleys and into the lowlands below until they spilled into the sea. The enormous force and quantity of water rushing through these small valleys caused widespread erosion and resulted in spectacular widening and deepening. This, in turn, produces a post-glacial phenomenon that belies the size of the valley today, since it is occupied by a much tamer river system.

There are many examples of these glacial spillways, but one of the more picturesque and imposing is the Columbia River Gorge that resulted from a gargantuan volume of water which was released from the unstable water body that lay trapped by the ice wall higher up in the mountains. Today, you can travel along the road that separates Oregon and Washington states, and enjoy the majestic views that are a legacy of the last ice age (but please, ... don't tell anybody I sent you).

Sometimes ice bergs break off the ice-front, called *snout*, and melt into the sea but when the melting occurs overland, then huge run-off is the result. Rivers do their own thing. They have upper, middle and lower courses and they erode transport and deposit silt in their own way. But when you have a large number of rivers washing out in front of the ice sheets, they distribute and sort all the material, called load, according to their ability. This includes detritus and debris that the ice picked up on its traverse over the terrain that it was scraping and scratching. How much material could this be? A lot!

Lake Michigan is the largest fresh water lake in the world and it was scoured out by a colossal lobe of ice that advanced and retreated, advanced and retreated, advanced again and retreated again, over many years as the temperatures fluctuated until finally it retreated back into Canada and left the present huge depression in its wake. Today it is filled with rainwater and is part of the Great Lakes.

Sometimes the rivers that run out of the front of the ice sheet spread great plains of outwash sand and gravel, and you know that sand and gravel does not make good soil for growing agricultural produce. This has a major impact on the livelihoods of people living in regions where the ice age left poor soils in its wake.

The Jutland peninsula of Denmark, in Europe, is a perfect example of this differential deposition pattern where the ice left a residue of one kind in front of its path and left a residue of another nature beneath the last melt. The lobe of ice occupied the eastern fringe of the peninsula and rivers flowed out west from in front of the melting ice sheet. These rivers deposited vast plains of useless outwash sands and gravels in the western region, while along the eastern margins the melting ice lobe dumped its load of detritus, called boulder clay, producing a rich loamy soil. The result is that farms in the eastern section of Jutland today are prosperous and modern, whereas the poorest and most meagre livelihoods have to be eked from the barren landscape to the west. All this from the ice age that passed-by twenty thousands years before.

In yet another instance of residual ice-age advantage, the lobe of ice rested on the North Germanic Plain, feeding torrents of muddy rivers that flowed east and west, according to the lie of the land in front of the ice sheet. Of course, as soon as the ice retreated and the water source was cut off the rivers dried up, leaving in their wakes distinct natural corridors that we were able to use in later years to carry major route-ways that link today's cities throughout the entire plain. These route-ways included railway lines, roads, canals and motorways and were important reasons for the rapid growth of Germany's industrial power in the last century.

Ice is heavy and it weighs down on the surface of the earth on which it sits. You already know that the earth is made up of a number of layers, the outermost being a thin crust that sits on a mantle and the innermost portions are molten and gaseous. All this makes for a rather plastic, liquid structure and the enormous weight of the ice causes the land to sink into the mantle below. You might say, "Yeah, the land sinks but that's OK, because the sea level drops by three hundred feet also, and nobody would notice the difference."

On the other hand, when the ice melted and flowed back into the sea, it raised the sea level back to its former position. The land, now relieved of the huge burden of ice on its back, didn't respond by springing back immediately. It slowly rose over the next ten thousand years and in many places it is still rising out of the water today. As the sea level rose in contrast to the surface of the land, it occasioned the drowning of low-lying areas and created islands, rias and fjords.

The sea promptly began to make its mark on the new landscape and cut notches in cliffs, laid out sandy beaches and wore away rock basins. Then over the next ten thousand years or so, as the land gradually regained its former level in relation to the sea, curious and exotic features were left high and dry above the present tidal process. We can see today wonderful examples of raised beaches, notched cliffs and wave-cut platforms in many places around our coasts. In the Baltic Sea, for instance, the coast line of Sweden receives approximately one foot of new land each year as that whole country continues to recover from the impacts of the recent ice age, a little more than ten thousand years ago.

This kind of fluctuation in land and sea level is called isostasy, and I'm sure that you are familiar with the term *isostatic recoil* for the re-emergence of the land to former levels in the aftermath of a great ice age.

There are many other after effects of the ice age also that we manage to live with today, some good and others not so good. The formation of the Great Lakes was in the long run a splendid natural phenomenon, since it was a major focus for transportation and greatly assisted the opening up of the vast reserves of raw materials and agricultural bases in this country in the last century. Today, it continues to flourish as one of the greatest sources of employment and habitation in the Midwest.

The continental ice sheets were such vast bulldozers that in their periodic advances and retreats they ground the rocks into flour, and when the ice finally melted, that rock-flour dried out to become a fine dust. The prevailing winds picked up the fine dust particles and blew them far and wide. Today, thick deposits of this fine silt called *Loess* are found in Kansas, the Ukraine, and Northern China, spawning rich agricultural pockets that sustain a vital component in the provisioning of each country's food basket.

How can you get a feel for what it is like during an ice age? How big is a continental ice sheet? These were questions that I thought about as I was reading my geography texts. But the year after my graduation, I was in for a surprise as I peered out the tiny window on a trans-polar flight on my way from London to Seattle. The sheer expanse and immensity of the ice cap was beyond my wildest imaginings prior to this, and I spent the best part of four hours staring in awe at the featureless terrain below, pleased that I was not down there in my blue jeans and sneakers.

The best way to get a feel for a major ice sheet is to imagine that you too are flying over the north pole, in a window seat. Incidentally, since you already know about geodesics and great circle routes, you will understand that many international flights go over the poles because this is the shortest route. Check your new globe!

As you look out your window over the vast expanse of wasteland all you will be able to see will be a white sea of ice and snow. You will be immediately struck by the immensity of this polar ice cap, and only then will you truly understand the dimensions of ice and the effectiveness and power of its work. From horizon to circular horizon you will marvel at the intimidating size of white frosting.

Upon closer observation you will notice that there appear to be mountains that stand out through the surface of that sea of ice, but they too are covered in the same white color. These are called Nunataks and they are high peaks of mountain ranges that protrude through the surrounding ice field. Notice that the ice flows downhill although it seems to not even move. You are shielded from its effect by the fact that you are at thirty thousand feet and everything seems still below. Even the clouds resemble an ethereal skyscape that seem to float-by in light wisps, without any connection to the earth beneath them. Can you name them?

While you are still in your plane flying over the polar ice cap you may see the place of transition, where the white landscape below gives way to a mixed land- and sea-scape. Here the blue water is dotted with huge blocks of ice that float away from the face of the towering ice-falls, marking the edge of the ice fields. What incredible scenic splendor - and good for you that you are not in a ship trying to navigate through the icebergs and ice-flows that caused such disasters as the Titanic and many others that are less well known.

You will also see that rivers of ice seem to flow through clearly defined valleys from the ice fields in the mountains above, sometimes stopping short of the sea and melting from their snouts into raging torrents that cascade down into the ocean below. This is the second form of ice that occurs sporadically throughout the earth, where ever the temperature is below freezing point for long periods of time.

Ice can flow like rivers also and we call them rivers of ice or simply glaciers. They normally begin high in the mountains and make their way to the sea or to warmer temperatures where they automatically melt, by occupying preglacial river valleys. The glacier does a similar type of work to a river since it erodes the terrain through which it moves, it transports the eroded debris, and deposits its load over a widespread area when the time is right.

As you look at the glacier it will seem to be a rigid frozen river, but nothing could be farther from reality. The glacier is constantly moving, crunching, groaning and working as it either advances or "retreats" in its course. If the production of snow up on the mountains is greater than the melt at the snout then the glacier advances, and if the melt is greater than the snowfall then the glacier is said to retreat. But all the while it is moving.

I camped on a glacier one time in Switzerland high up near the *Aiguille de Tre la Tete*, and I will always remember the noise that emanated from the 'sleeping giant' as I made my way over the crevasses, seracs and ice falls up to the summit. I could hear water flowing at certain times somewhere inside the glacier, and every so often, I heard loud booms when avalanches roared down the slopes up ahead. A friend who was climbing with me complained constantly about the eerie sounds, and once I heard him mutter under his breadth: "This darn thing is alive!"

The glacier will be fed by falling snow up at the highest elevations and it will typically end somewhere down the valley below in a cascading, tumbling, cacophonous discharge from the snout where it melts. To the casual onlooker the glacier appears to be always at the same position in the valley, but remember that it is constantly being fed by acres of snow from above and it continually outpours raging, crashing rivers at the end.

If more snow than usual falls in any particular winter, or if it stays colder longer, the glacier will advance a few feet down the valley. At the same time, if the summer is longer or warmer than usual the glacier will retreat up the valley a-ways. Overall, however, the glacier will occupy the same position in the valley, year after year, with little or no appreciable difference evident to the non-scientist. I remember an entrepreneurial hotelier in Switzerland who built his pension on a very picturesque valley in full view of the local glacier and naturally, he called it "Glacier View Hotel".

The glacier wasn't very cooperative to his scheme, however, and over the next fifteen years the ice continued to retreat until finally, it no longer was visible from the verandahs of the exclusive rooms. He was stretched on the horns of a dilemma. Should he rename his pension "Used to be Glacier View Hotel"? Should he go with a more traditional Tyrolean theme, like *"Alpen Haus"*, or should he simply take a class in remedial geography?

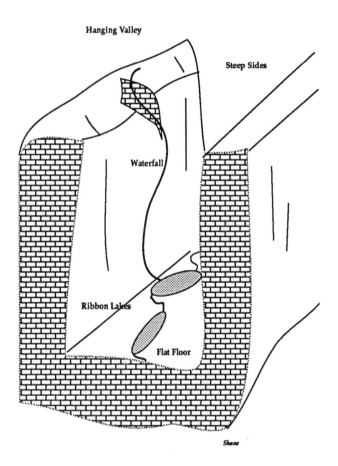

Fig. 14.2 Typical Glaciated U-shaped Valley

The glacier begins up in the mountains in a snow filled depression that is referred to by several names, (tarn, corrie, cirque) each meaning the same thing. Today, where the ice has melted you can find these depressions filled with fresh water to make some of the most beautiful mountain lakes on the planet. These corrie lakes or tarns are places to which I love to go hiking, armed with my camera and a packed lunch. Other people often fish these lakes but I'm too lazy for that, unless I could get you to lug my gear up into the valley for me. Then I wouldn't want to go because it would be too crowded. I live for the solitude of a lone glaciated valley where I can listen to the solitary cry of the raven as it soars unencumbered over its empty vastness.

Sometimes, on a mountain there will be a corrie glacier on the north face and another one on the east face and another one on the southwest face, each eating away at the mountain back-to-back. Finally, when the glacier retreats and melts, there will remain in place of these erosional corries three very steep-sided cliffs surrounding one pyramidal peak with knife edged ridges, called aretes, leading from the summit. The Matterhorn, in Switzerland, was formed in this fashion and you can get splendid views of the corrie backwalls, the pyramidal peak, and the arete from the Hornli Hut or from the wonderful train journey up the valley.

The glacier erodes, transports and deposits just as any river does, but with a different technique and with different tools. Think of the river of ice as a giant bulldozer that widens and deepens existing river valleys and essentially changes their shapes from the traditional V-shaped to the glacial U-shaped valley. From this description, you can accept that V-shaped valleys are formed by rivers and when you see a U-shaped valley you can deduce that it was chiselled by a glacier.

Erosion happens by a combination of frost action, sometimes called freeze/thaw action and plucking. Continual freezing and thawing causes rocks to split and crumble. Water seeps into the cracks, freezes, expands and more freezing causes them to spread even more and eventually split. When the frozen water becomes attached to the glacier the entire rock is "plucked" from its position as the glacier advances and retreats. The valley floor becomes polished and scoured as the glacier moves over it.

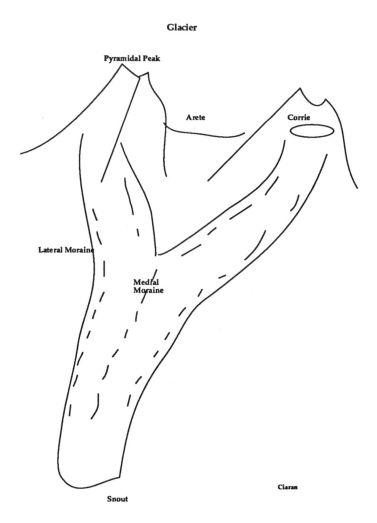

Fig. 14.3 Glaciers in Action

When the glacier melts and the valley again lays exposed to the elements we can see the remnants of the ice age. There are rock formations called *roches moutonnees* after the French for wig (the kind that judges used to wear in France in the nineteen hundreds.) This peculiar formation transpired when the ice scoured one side of the rock and plucked at the other side as the ice progressed down the valley. You can tell the direction of the ice flow by looking at the roche moutonnee and viewing the polished side and the plucked side.

The typical glaciated landscape is a result of a combination of the workings - erosion and deposition - of the ice and the accompanying rivers that emanated from underneath and from in front. Drumlins, eskers, moraines are a few of these common features and they are easy to pick out when you are having lunch at your favorite corrie lake waiting for the raven to call its piercing cry.

Moraine is a name for the load that the glacier transports on top or in the middle or underneath. Rocks and debris that fall on to the glacier surface because of frost action, bedrock and soil that is plucked from the floor as the ice moves over it, and any material that falls into crevasses because of avalanches or any other method all make up moraine. Typically, a glacier will have two clearly visible lateral moraines on the sides and sometimes, when two glaciers merge from tributary valleys, the two inside lateral moraines coalesce to form a medial moraine. That sounds very complicated, but in fact it is the most logical thing in the world - look at the diagram (Fig. 14.3).

The glacier moves a great deal of morainic material and all this is dumped at the end to be sorted and graded by the rivers in front of the ice. The moraines that are left sporadically by the ice, as the glacier retreats over a number of years, are called recessional moraines: the last moraine that remains after the glacier has finally retreated, and melted right back to the corrie, is called a terminal moraine. Lakes are formed sometimes by rainwater that fills glacial depressions resulting from erosional features like plucking and glacial scouring, but also from morainic damned hollows that trap the runoff.

Rivers, formed from meltwater underneath the glacier, tunnel out from in front and are fed with debris and morainic material from inside the ice. These rivers can run for great distances under the ice and create their own beds and flow patterns. When the ice is finally melted the river disappears and in its stead, plainly visible, is the esker that snakes across the land imaging the path of the glacier and the river that flowed underneath. These eskers were usually higher than the surrounding countryside and were later used for building routeways and especially railway lines.

Glaciers do not like to negotiate corners too well, so instead of trying to go around them, they cut them off or truncate them. These corners are called spurs in geographical terminology and the feature that is most noticeable when you look up, or down, a glaciated U-shaped valley is how singularly steep and straight the sides are. This is because the interlocking spurs that you associate with a normal river valley have been truncated, creating very picturesque scenery, but there is more.

Very often you will be able to pick out beautiful and spectacular waterfalls tumbling over the rim of the steep sides, rushing to reach one of the placid lakes that usually dot the floor of the valley. These are called hanging valleys and they were formed when the glacier filled the existing valley and a little tributary flowed onto it from above. When the glacier melted the tributary now had to tumble to the floor below and resulted in the delightful visual scape. Meanwhile the many charming lakes that string out along the valley floor in a formation that has often been referred to as 'Pater Noster' because they resemble the beads in a rosary, are called ribbon lakes.

Yosemite valley in California, provides a breath-taking example of this enchanting scenery, but be aware that there are as many hanging valleys, ribbon lakes and cascading waterfalls as there are glaciated U-shaped valleys on the earth, and you can find your own unspoilt, pristine spot in seclusion with a little inspired map reading, and a lot of energetic footwork through recessional moraines. Don't despair at the prospect of some arduous foot slogging, because after all, it is for this very reason places, like these that epitomize rugged splendor, remain unspoilt, pristine and secluded and a joy for all of us.

![Chapter heading bar]

Chapter Fifteen:

Vulcanism

I read recently about a Boeing 747 that was cruising at an altitude of thirty five thousand feet over the Malayan Peninsula when suddenly the four engines died. The stricken jet immediately went into a dive and plummeted earthward, while the captain and crew frantically grappled with the controls in an effort to restart the engines. The passengers, those who were not secured by seat belts, were hurled about in the interior and panic was widespread.

The plane dropped from thirty-five thousand feet to about three-thousand feet before the captain was able to get the engines to power-up again and regain control of the doomed craft. Then, as the confusion began to ebb, they came to realize that they had flown into a thick cloud of ash and dust which had been catapulted into the atmosphere by an erupting volcano. The thick pall of dust had choked and clogged the engines occasioning the sudden and unexpected loss of power, and the near disaster. For the crew and passengers in this flight the reality and the devastation that can result from a volcano were very real and it brought home to everybody that there were vast forces present in our earth with which we have to contend and understand.

You can truly feel a sense of the tremendous power and immensity of the universe when you see a volcano erupting, even from the comfort of your own living room on TV. At times like these we are reminded of our universe's fiery beginnings and the incredible energy that is stored within our own planet. How does a volcano erupt and what makes it do the things that we recognize today as part of vulcanism?

Volcano

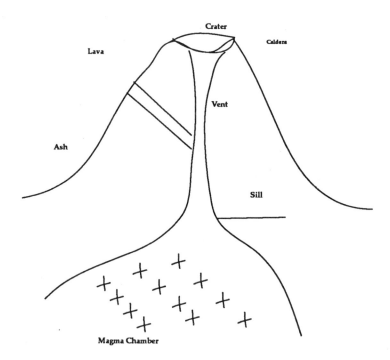

Fig. 15.1 Volcano

When we think of volcanoes the first thing that springs to mind is a crater, atop a conical mountain, with perhaps snow on the upper reaches. But there is a great deal more to learn about vulcanism and the effects of volcanoes on the landscape and on our everyday lives. If you already live under a volcano then you know what I'm talking about. But if you are like the millions of other people, whose only glimpse of a volcano is sporadically on the news and in occasional National Geographic documentaries, then you can glean a good deal of information from an article such as this. Did you know that early man feared and worshipped volcanoes and they named them after the God of fire - Vulcan?

How do volcanoes erupt and what causes the catastrophic build-up of pressure that results in the fiery cataclysm? There are many theories pertaining to the rationale for the intense heat at the center of our planet. The question still remains, "Is the earth a globule of solar gas that is still cooling down as a result of its present orbit around the sun, whence it was pulled by a cosmic occurrence?" Or, "Is the earth being heated by nuclear forces from within that we have yet to discover?"

For whatever reasons, we already know that the core of our planet is gaseous and under tremendous pressure and intense heat. Outside the gaseous core is a layer of molten material called the astenosphere, which is fiery hot and also pressurized. The entire concoction is contained in a tenuous shell, called the earth's crust. Another name for the earth's crust is lithosphere, from the Latin word lithus - meaning stone, because it is comprised of rock with thicknesses that vary from upwards of twenty-five miles to only eight miles in places.

The earth is not one rigid, uniform block, but a collection of clearly defined tectonic plates that float on the liquefied infrastructure beneath. The tectonic plate theory is relatively new and does an adequate job of describing how the differing components of the planet are held together and how they seem to move by convection currents.

A much older theory is incorporated in the tectonic plate theory to describe continental drift. Continental drift is easily visualized with the aid of a crayon and a balloon. Draw a rough sketch of the earth on the balloon. Inflate the balloon and watch the continents drift apart. Inflate it even more and they move very far apart. Now let some air out and watch them fit back to an original pattern. Everybody knows that the similarities of the coastlines of south America and Africa are such that it is easy to imagine them closer at one time. Also the coastline of Madagascar and Africa and the coastline of Australia inverted and Antarctica all fit nicely together. It is also very easy to reconcile the geographical shapes of north east Africa that seem to be moving away from each other daily.

The important fact to realize when you are attempting to come to grips with tectonic plate theory is that the plates need not be contiguous with the coastlines. Only then will you be able to comprehend that the Pacific plate, which is comprised mostly of ocean, also contains a sliver of Washington, a portion of Oregon and a segment of the Californian coastal plane. Since the North American Plate is moving in one direction and the adjacent Pacific Plate is moving in the opposite direction it is easy to see how they continuously crunch against one another, wreaking tumultuous havoc in places that sit on the edge like San Francisco.

This is just one particular zone of confrontation, and it has created the infamous 'San Andreas Fault'. We are all familiar with this fault because there are so many impressive volcanoes sitting on it, including Mt. St. Helens that erupted recently. When the plates collide, one is forced under the other, in what is known as the subduction zone. In each zone, the stress and fracture lines present in the crust enable the activity of vulcanism since the tremendous pressure and heat can penetrate and manipulate these weaknesses. This forces the rock structure into the magma chamber increasing the pressure and causes stress fractures to erupt into explosive volcanoes. There are volcanoes evident above the subduction zones of all plate edges and particularly in what is called the "Pacific Ring of Fire." Finally, when the temperature and pressure is greater than the retaining power of the crust an eruption occurs.

At other places there are zones of divergence where new rocks are formed. This occurs at fracture lines where two adjacent plates are pulling away from each other, as is taking place on the east African plains today. The African rift valley originates at the north and extends three thousand miles creating some of the most majestic scenery and uninhabitable regions of this planet.

Some divergent activity often happens unseen at the bottom of the sea, called simply, *ocean-floor-spreading*. This results in a number of interesting phenomena that weren't discovered until surveying methods and oceanographical processes were aided by modern technology.

The first of these was the discovery of mid oceanic ridges where fresh lava was upwelling from fault lines and immediately cooling to form new rocks. A classic example is in the Mid-Atlantic, where a major oceanic ridge runs from Iceland to the south Atlantic. For practical purposes, you might think that the ocean is at its deepest at its center. However, because of the existence of the mid oceanic ridge, there are two deep troughs, one off each coast, and the center is shallower than either of the channels. Take a peek at your atlas and check the submarine contours, or simply look at the color patterns. Another result is that newest rock is formed at the mid oceanic ridges and here, scientists can get a glimpse of the forces and elements that fashion the earth.

In places, the eruptions are so numerous and so powerful that they have created significant undersea mountains, where the summits protrude just above the surface to show as volcanic islands. Some, like the Hawaiian Islands, the Icelandic Islands, and the South Sea Islands were formed in this way.

Rock that is maintained at formidable temperatures changes to a liquid, a red hot molten soup called magma. The soup analogy is a good one because when soup cools down a thin crust forms on the outside and the inside can still be hot and liquid. The lithosphere is melted down in places and here the magma builds up in what are called huge magma chambers.

Mid-oceanic Ridge

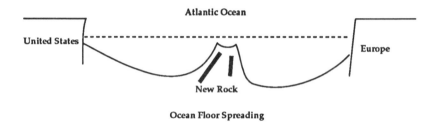

Ocean Floor Spreading

Fig. 15.2 Mid-oceanic Ridge

Magma cools down when it reaches the surface of the earth or when it comes in contact with cold air or water. When the magma cools it forms different kinds of rocks. The rate at which the magma cools can determine the nature of the rock that results.

The super-heated mantle exploits endemic cracks and fissures, eventually penetrating within a few miles of the earths crust. The pressure and temperature continue to build until finally the weaknesses in the earth's crust yield to the superior forces and the magma often explodes through to the surface in a spectacular display.

Sometimes, there is no spectacular display, no bombastic cloud of gas or steam and no torrent of red-hot lava, just a quiet surge of molten substance spreading slowly over the land. Lava is the name that is given to magma when it reaches the surface and some of the pent-up gases escape or are emitted.

The magma chamber reaches through to the surface by means of a central pipe or vent, and the gases and lava are released in successive explosions and fiery arrays. Over a number of eruptions or during a prolonged eruption the lava will build up around the vent to form a cinder cone or a conical mountain. The central vent will form a crater on top and sometimes this will fill with water. When the volcano hasn't erupted for a few hundred years it is said to be inactive or dormant, and when it hasn't groaned into life for thousands of years it is said to be extinct, but many people think that volcanoes rarely become really extinct.

There is so much heat and pressure associated with an eruption that all the surrounding rocks become liquefied and fault lines and cracks are easily exploited so that secondary vents often appear in the sides of the main cone and new lava flows issue forth. Sometimes also, the weight of the cone-shaped mountain causes the entire summit to sink into the super-heated base and a very distinctive cone-within-a-cone results. This is called a *Caldera* and there are many fine examples of calderas in the volcanic regions of the earth, but specifically mount Kilimanjaro, in East Africa, is a fine example.

Often when the heat and pressure builds up again and another explosion takes place, the top of the mountain can be blown off and dust and ash will enter the higher reaches of the atmosphere. When this happens as occurred in Krakatoa in 1910, the dust circulates high above the earth in the stratosphere and we experience very vivid sunsets for a number of years. In 1991, wonderful sunsets were apparent after Pinatube's eruption because there was so much dust in the atmosphere. These particles in the upper atmosphere can also block out rays of the sun and have an adverse effect on our climate.

But all effects of volcanoes are not adverse, detrimental to our health and a danger to our well being. There are various rich and diverse minerals present in volcanoes that are valuable to man and great quantities of sulphur are produced at times of eruption. In addition, the ash and rock that are deposited over wide areas of the earth during eruptions weather to form rich loamy soils and the basis for a diverse agricultural industry later.

In parts of China, where volcanic activity has been prevalent for thousands of years, the people have learned to take advantage of the vast energy that is generated by the upwellings of molten magma. Water that infiltrates into the magma chamber becomes steam and fumaroles and geysers are widespread. I'm sure that you have seen pictures of *Old Faithful* - the geyser that erupts with precision regularity at Yellowstone National Park. This is simply a penetration of water into the magma chamber below where the intense pressure is built up until it explodes into the spectacle that so many tourists love to see.

Energy that is stored in the volcanic activity and can be put to work for man is called *geo-thermal energy*. This is the kind of thing that the Chinese discovered and put to use to heat their houses and to generate electricity. We must remember that there is incredible energy locked up in our earth, just below the surface. All we need do is figure out how to control it. This might seem an over-simplification, especially if you witnessed the eruption of Mount St. Helens when it recently showed the world just how destructive nature can be. Many people, who now live along the western fringes of the United states watch with interest the reports about bulges swelling-up on the sides of their favorite scenic vistas, like Mount Rainier, or Mount Hood or any of the umpteen dormant or extinct volcanoes that line the Cascade Range from Canada to Mexico.

Me! I stopped walking near volcanic peaks when jets were scheduled to fly over them, you can't be too sure, nowadays.

Chapter Sixteen:

Deserts

Desert Storm brought the vastness and the emptiness of the desert into all our living rooms. Up to that, many of us had never even seen a desert. Of course we knew that deserts happen to be hot and dry where snakes and cacti proliferate. We also knew that they are sandy, except beside drinking holes, called oases where camels live with their nomadic owners.... or is it the other way around. And that is that! Should we know more than this about deserts? How come there is so much oil in the deserts and all these oil sheiks are so wealthy? Has this got anything to do with deserts? Many people have images of deserts as areas of sand dunes (see figure 16.1), but most deserts are actually barren rocky spreads of desolation. And I'm sure that there is as much oil under the ocean as there is under the desert skies. So there!

About one third of the earth's land surface is occupied by desert and semi-desert. These regions are arid, that is, they receive very little rainfall per annum. Even when there is some rainfall, it is of little use, since there are such extremes of heat that it immediately evaporates. If you look at a world map you will notice some interesting facts about the location of the world's large desert tracts. Primarily, the deserts are all in the same latitudes, roughly twenty-three degrees north and south of the equator, and secondly, they are each situated on the western fringe of a continent. There are good reasons for their predictable locations and we can examine them. A knowledge of these reasons makes it easy to understand the concepts associated with the advance of desertification and we can explore a simple cause and effect operation.

A Typical Sand Dune

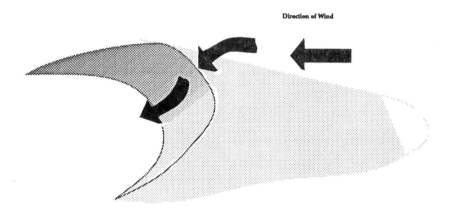

Direction of Wind

Shane O'Mahony

Fig. 16.1 The Barkhan Dune

The main cause of desertification is the perennial lack of precipitation. We already know that rainfall occurs from clouds and that clouds are but air masses that are rising, condensing and releasing the moisture contained therein. This is simple. When air rises it cools and as it cools it condenses and loses its ability to hold moisture. Ergo rain! Now don't mix ergo with erg!

You know that the sun shines continuously on the equatorial lands, between the Tropic of Cancer in the northern hemisphere and the Tropic of Capricorn in the southern hemisphere. But this is not exactly where the deserts are. Hot air rises, cools and condenses bringing equatorial rain in these regions, with associated lush foliage and dense jungle. The air that is heated by the hot sun rises and moves away from the equator. It is replaced by fresh cooler air that flows in from all sides.

The deserts aren't exactly on the equator.

The displaced equatorial air is now losing its heat, as it moves out. As it does so it descends and replaces, in turn, the air from around the tropics of Cancer and Capricorn. So this air is in fact descending from higher up and expanding as it begins to heat up. Its capacity to absorb moisture is increasing and that is the opposite to rain, but the matter is compounded by the fact that since it moves over land there isn't much moisture to pick up anyway.

The results are predictable. On the west-sides of large landmasses near the tropics, north and south of the equator, deserts flourish. Some of the more notable desert regions are; the Sahara, the Atacama, the Arabian, the Great Australian, the Kalahari, and the Californian. You could locate these exotic places in your atlas and verify, for yourself, their positions with respect to the equator and their situation on the continent. See that they are each roughly twenty-three-and-a-half degrees either north or south of the equator, and that they are each on the west side of a continental landmass. You may well ask why west, why not in the middle or even on the eastern fringes? This latter phenomenon occurs because the winds are deflected from east to west due to the direction of rotation of the planet earth. Coriolis force... but you already knew that!

While the natural order of things may have caused deserts to form, the addition of man and his eternal search for progress has not always helped stabilize their spreading, and in fact, it is estimated that millions of acres of land are destroyed yearly as a result of man's misuse of the elements. Desertification continues to expand because of the habitual loss of fertile soil on the perimeters of deserts. The spreading of deserts is causing increasing concern in a world whose population is expanding, since it reduces the area of land available for food production. Over-grazing of livestock is the chief cause of the loss, while unscientific farming practices and destruction of trees contributes greatly to the grim picture.

Reclaiming the desert from the elements is a slow and tedious process and some attempts have already been made. Educating our farmers to utilize the land with enlightenment, which in the long run benefits each of us since in reality it is all our land, has always been a worthwhile project. Invariably, it results in improved standards of living while still maintaining a sane use of our natural resources. When I say 'we', 'us' and 'our' I'm speaking to the big picture, because we must all take responsibility equally for the safe preservation of our planet into the future. By using improved agricultural methods, and by placing limits on the amount of farm animals that graze the land, some initial steps can be taken to help curtail the spread of deserts and conserve our agricultural lands. Other practical measures that we can take include the planting of long-rooted grasses and trees, as wind shelters. These actions help stabilize the desertification process and halt the migration of the sand dunes over bountiful arable land.

This shouldn't be a surprise to anybody. It's not like we haven't heard of these breakthroughs before. Remember *The Grapes of Wrath*? Many and salutary were the lessons that were learned in the aftermath of that destruction. A classic example of man's ability to wreak immense havoc on the landscape and then endure the insufferable consequences, was witnessed in the nineteen thirties, in what came to be known as the Dust Bowl. Over fifty-million acres of topsoil were blown away and the damage was so severe that the effects are still being felt and will continue to be felt for many generations. The states of Colorado, Kansas, New Mexico, Oklahoma and Texas in the southern Great Plains were affected and thousands of bankrupt farmers were forced to abandon their way of life and seek new opportunities in other parts of the country, especially in California.

This episode, attesting to man's travail against the elements and his relentless attempts to temper mother nature, was captured with poignancy, by John Steinbeck, in his haunting story. All school children should read this book, both for the scholastic and linguistic treat, but also to study it from the geographic and ecological viewpoint since it chronicles man's ignorance and mismanagement in relation to his environment.

We learn again and again from our mismanagement of nature's resources.

The states of the Great Plains are natural grasslands. Man ploughed some of these lands converting them into wheat-producing fields, but did so in blissful ignorance of the hazards they were imminently creating. This left the topsoil inadequately protected from nature's elements - the sun and the wind. In addition, the remaining grasslands were destroyed by over-grazing of too many livestock. Aeolian forces are immensely powerful, with capabilities to erode, polish and sculpt deflation hollows and other dramatic ventifacts, such as inselbergs, yardangs, and zeugens. The sun shone and the winds came and the land was ruined. Too late to arrest it, the farmers were powerless to halt the disaster. Help came too late also. It is said that while tons and tons of topsoil were being borne aloft in the air, the politicians who were 'deflation debating' what to do about the destruction, had to shut the windows in their offices in Washington DC to keep the sand off their papers.

Nature is relentless in its activity, and man must be ever vigilant not to upset the delicate balance that has been established since time immemorial on our planet. In the aftermath of the big wind sound farming practices were implemented, shelter belts were strategically positioned to control the wind force, and life continued as before. But unfortunately, man forgot his lesson and as soon as the crisis was perceived to be at an end, old land utilization practices again began to creep in. Once more we were abruptly reminded of our place in the cosmic world, when storms in the mid fifties and again in the seventies wreaked havoc for the second and, some say, third time. Could it happen again? Are we responsible?

> **Man's ruination of the environment. Of course it can happen today! It is happening right now.**

It may have a different name, and it may be in a different location but the underlying crisis is still an issue - man and his environment, versus enlightened man and his planet. Today, it may be the utilization of Native American grounds, which are not restricted by control regulations, for the disposal of industrial, domestic and hazardous waste. Or it may boil down to the conflict between the spotted owl and the unemployed lumberjack, but believe me we are still talking about the dust bowl and man's ability to wreak havoc by short sighted exploitation of natural resources for personal security and short-term, pecuniary gain.

Deserts are spectacular examples of nature's diversity. They are a real thrill to explore and they are rich in a kind of life that is unusual to most people. By studying deserts we enrich our view of our planet and, in addition, we get to view in a new light some of our resources which we tend to take for granted. We can see how valuable natural resources like water and agricultural land are, after we spend some time in the desert even if only figuratively. The survival instincts of the desert inhabitants are wonderful testimony to the amazing ability of nature at adapting and proliferating. It is good for us to learn how man can adapt to the harsher elements of the planet upon which he is a traveler.

Chapter Seventeen:

Oceans

I must go down to the sea again,
To the lonely sea and the sky;
And all I ask is a tall ship
And a star to steer her by.

Masefield

The earth is made up of almost seventy percent oceans. Scientists believe that all life sprung from the sea at one time. Our planet is the only planet that we know of, where water and oceans wash the shores. This is because we are situated at exactly the right distance from the sun, so that minerals and atoms needed to form water are in abundance. The world is really one large ocean, broken here and there by islands that we call continents. This great body of water is vast, and it is deep; up to six miles deep in places. Can you point out the deepest place in the ocean? It is the Mariana Trench, south west of Guam, reaching down to thirty-six thousand, one hundred and ninety-eight feet.

The Mid Atlantic Ridge is ten thousand miles in length from top to bottom.

The continents are bordered in places by a gently sloping shallow stretch called the *Continental Shelf* that is teeming with an amazing variety of animals and marine life. It also contains vast quantities of mineral wealth and fossil fuel, like oil for example. In other places the continents drop off precipitously into the murky deep, where there are great bodies of upwelling cold water. Such a cold current causes coastal fog banks as it meets the warmer air from the adjacent landmass. You may be familiar with the characteristic fogs that plague the San Francisco Bay Area and pre-empted Mark Twain's famous lines.

A cold upwelling of water from the deep produces cool air that mixes with warm Californian desert air, to form thick banks of fog. The result is that much of the coastal plains are prevented from viewing the sun for much of the summer and driving can be hazardous in the valley. An interesting sideline resulted from the existence of fog over the bay at the time that discovery ships were plying the coast trying to uncover the secrets of the shore. San Francisco Bay was one of the last places to be explored because it was so often fog-bound and the explorers stayed out to sea in safe waters, passing merrily by.

Oceans have fascinated man from the very beginning and today they are still impressive so that seascapes inspire paintings, poetry and prose, while adventuresome folks test their skills and their ambitions by taking-on the might and power of the sea.

I grew up around sailboats. I can never remember a time when we didn't have access to the sea and a sturdy boat in which to explore and forage among the islands and inlets in my native land. I learned to love and respect the sea from the elder mariners who used to sit on the quay walls and talk about tall ships and stars, but who always had good advice about the weather, currents and the tides. It wasn't until many years later, when I chanced to go crab fishing in the Bering Sea in the winter season that I truly came to understand the power and awesome might of the sea.

We were wearing survival suits and were confident that we could withstand the cold, but I was not prepared for what happened. I never saw the rogue wave that crashed in over the transom, as I grappled with the gear. According to the captain who saw all, and told me later, I was picked up and never touched the ship as it swished me bodily overboard, dumping me into the arctic foam.

My first instinct, when the icy waters gripped my being, was to grasp at the net that was in the water beside me. This turned out to be a bad mistake since I was immediately sucked down by the cable. I scrambled, gasping, and panic-stricken to the surface once more and managed to waste a great deal of energy with wild, flailing actions. With horror I realized that the ship was gone. Gone! The stark severity of the situation stunned me even more than the frigid ocean rollers.

My ship had come in, and was gone.

The ship was gone! Chugging off through the crashing waves, laying pots for lobsters. Are they mad? Look at me, freezing to death, drowning in this mountainous icy grave. Shouting was no good. I couldn't even hear myself in the drama and the emptiness. I remembered my training and lay on my back in the survival suit, to conserve my energy and maintain any heat that remained. Heat! What a joke. Had anybody seen me go overboard?

They tell me that it was only seven minutes, but I swear that it was more, when the bow loomed out of nowhere and I was rescued. The mate cut the hawsers with the acetylene torch, dropping the net (and $500,000) into the deep, turned the vessel and retraced his course to where I lay motionless. We returned to port empty-handed but full of awe and with a fearsome respect for the Bering Sea.

Since then I have stayed away from crabs and lobsters, but I still love to sail, especially when the sun shines and the breeze comes in from the south west. There is so much to learn about the sea that you could spend your entire life working at it, and still some. A good place to start is at the beach, where you can enjoy the fresh air, the sunshine on the water and the rhythm of the waves breaking over the sand and pebble-shore.

What are tides? There are high-tides and low-tides, spring-tides and neap-tides. What does all this mean? And how do tides effect us in our everyday lives? Do we even need to know? A quick glance at the planetary ballet will reveal some of the answers to these questions.

The earth rotates on its own axis once every twenty-four hours. At the same time, it revolves about the sun in a huge ellipse that takes a little over three-hundred-and-sixty-five days. It remains locked in its solar orbit because of the effect of gravitation. Meanwhile the earth has a moon that circles about it, and is maintained in this path because of its closeness to and the resulting gravity that attracts it to the earth.

Twice a day, the sun, moon and earth are so aligned that their cumulative or scattered gravitational pull effects the water in the oceans and we experience high and low tides. Periodically, the earth, moon and sun are lined up so that their gravitational effects are cumulative. This causes a particularly high tide which we call Spring-tide, and at other times the sun's, moon's, and earth's gravity are working against each other so that a particularly low tide is the result. This is known as a Neap-tide. You cannot have high tide at all places on the earth at the same time. A high tide here will mean a correspondingly low tide there. The section of coastline that is effected by tides usually is clearly visible, because it will have muddy water-marks and seaweed remains that leave prints on the rocks and beaches. This area is called the tidal reaches and will be of interest to you if you own a moorings and have your boat tied up in a tidal harbor.

Tidal scouring is a good thing for many harbors because the ebbing flow helps flush out the channel and keep it clear of any debris or pollutants. Some harbors have sand-bars across their estuaries and navigators prefer to enter these channels when the tide is full and there is sufficient depth for ships of large draft.

The abyss is made up of deep gorges and canyons.

I have always thought it peculiar that the high tide occurs not only in front of the gravitational pull below the moon, but also at the opposite end of the earth. That was hard for me to understand, so I asked my teenage scholar. Why does the moon effect the tides on the opposite side of the earth? He hadn't even heard about it. What do they do all day in school, I wondered!

I ruminated over the problem for a few days until I had an opportunity to go to the library. My friendly neighborhood librarian will surely know the answer! It caused quite a hubbub when I presented my investigation on Saturday morning and there were no less than three librarians anxious to learn the answer to such a perplexing question. There was a general sigh of satisfaction when we finally unearthed the reasons. It made perfect sense to me. I was so pleased to rid my mind of that perplexing question. It even made sense to my high-school 'scholar'.

As soon as I had him in a learning mode I thought I should capitalize on it, and introduce a few new concepts to him. At the risk of overload I asked him about waves. How are waves formed and what drives them? How come some waves are so small and others are monstrous like those ones he sees on TV at the surfing 'Pipeline' championships! If as he rightly infers, waves are the result of wind blowing the surface water, how do you explain the lack of waves on some windy days at the ocean?

These are good questions and are perfect to test the affective agility of a teenager and great to stretch his cognitive skills, while he grapples with the mental extrapolations associated with physical observations.

Waves are caused by the friction of the wind on the surface waters, but only if the wind and the water are traveling in the same direction. So that explains why there will be a flattening effect if the wind is blowing against the water motion, and no waves will occur. You may stand at the edge of the ocean on a windy, blustering day and wonder why there are no waves. Now you know! The wind and the water direction need to be coordinated.

A wave resembles a giant wheel and it moves forward in a circular motion through the open sea, with no interference from land masses. Waves need large bodies of water and a strong wind to be able to attain goodly sizes. Take for instance the Atlantic Ocean between the east coast of North America and the western limits of Europe. There is a large body of water here and there is a prevailing southwesterly wind that blows across that body of water. Waves are generated and forge ahead with great momentum unimpeded by any land mass. A stretch of water like this is called a fetch, where the wind can whip up the sea into a frenzy of activity.

What happens when the wave approaches a land mass? This is the part with which most of us have familiarity, since very few of us are lucky enough to spend time in the mid-Atlantic fighting rollers and quartering seas. Look at the wave as a huge wheel. It is easy to visualize the top because we can see it and its shape above the surface of the ocean. But there is another part of the wave which is hard to see beneath the surface and it completes the circular motion. You may have felt its power when you went swimming and got knocked over by the drag of the water as it rushed about in a pre-conceived arrangement.

When the wave is in the open sea it has plenty of room to make its revolution, but as it nears the land mass, the bottom of the circle 'trips' on the sea floor and the circle is broken. We call these waves *breakers* and we can see the white foam on the tops of the waves indicating that the circle has broken and the wave is about to collapse. Of course the wave crashes with all its might onto the land and causes destruction and splendor.

Wave Action

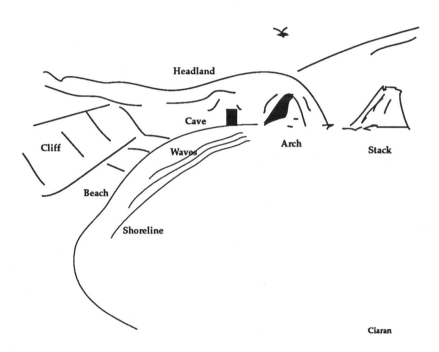

Fig. 17.1 Seascape

    The wave action erodes and sorts material, making a flat section at the edge of the sea, known as a wave-cut platform. The erosive and depositional actions of the waves cause beaches to build-up, sometimes made of pebbles and sometimes made of sand. The wave also known as swash, rushes on-shore with rhythmic cadence, and then the backwash slips down the slope to rejoin the next oncoming surge. This is easy to understand. The breaker crashes onto the beach with tremendous power and runs up the beach until it dissipates, but returns to the sea urged on by the pull of gravity. The water has more energy on its shoreward run and because of that a peculiar thing happens to the beach profile.

Have you noticed that all the pebbles and rocks that you pick up on the beach are smooth and even flat. Why is this? Can it have something to do with the eternal movement of the waves rubbing the pebbles against the sea-floor and against each other? This is called attrition, when stones and rocks are smoothed and rounded by hydraulic action and the action of abrasion against other stones and rocks. It is the same attrition that happens in the beds of rivers, as you may recall from earlier. There is a simple rule of thumb that we use to summarize the observation of rocks and stones. If the pebble is angular, then it is a result of frost action, and if the pebble is smooth it is more likely the result of hydraulic action by the sea or by a river.

The shoreline is sculpted and created by the action of the waves and the other elements so that it can be a very picturesque and scenic place, especially in juxtaposition with the rhythmic cadences of the sea and the wildlife.

Beyond the beach the waves cut a notch and eventually a cliff can be formed. Depending on the hardness of the existing rocks, some excitingly spectacular features can result from the hydraulic action of the waves and especially the storms, when there is a great concentration of the power of the sea. Sea caves occur from differential erosion when the erosive power of the sea exploits weaker strata in the rocks and wears away large hollows. Sometimes two caves eating back-to-back will coalesce to form a sea-arch and these are a pleasure to behold.

Later, when the arch collapses with time a sea-stack will remain like a small island on its own, home for the gannets and sea terns that soar and scream around the cliffs and out over the bay. Pocket beaches might form to one side of the bay as a result of erosion to the less resistant bands of rocks and you can find wonderful secluded niches to sun bathe and picnic.

If you look at some beaches, especially cuspate or half-horseshoe beaches, you will notice that all the sand is at one end and all the pebbles and rocks are at the other end. This occurs because of a phenomenon called *long-shore drift* and is a direct response to the action of the waves that we just described. You can prove this for yourself with a simple experiment.

Watch a piece of flotsam from the beach and observe its motion carefully in the swash and backwash. You will notice that when the sea surges forward it does so at an angle to the beach, usually because of the configuration of the shoreline which is rarely perpendicular to the wave action. This causes the flotsam to lurch forward and sideways a little but as it retreats with the outgoing wash, under the principles of gravity, it simply drifts backwards in a straight line.

Next time it lunges forward again it does so at an angle, only to retreat in a straight line back into the sea. Watch this continue for a time and you will perceive that the floating object has drifted down the beach quite a ways. This is what would happen to unsuspecting swimmers also, if they failed to notice how they were being carried down the beach by the action of the waves. When finally they come ashore after their swim, they find they are nowhere near the place where they had set out earlier.

Long-shore drift has a greater impact on our coastlines than we may suspect. Especially when you add this kind of action to the cumulative effect of rivers and ice and wind, and so on. Remember how we summarized rivers by saying they eroded the sides and channel bed, transported the resulting load, and deposited the silt in the sea. Well, here we are in the sea and all this silt is being deposited by rivers from all over. Take a look at your atlas and observe the west coast of North America from Alaska to Mexico. Count how many rivers flow into the Pacific from the Cascade Range and the Rocky Mountains. The larger rivers include the Yukon, the Athabasca, the Snake, the Columbia, the San Joaquin, the Sacramento, and the Colorado. These rivers are no small streams. Together with the hundreds of smaller rivers, they deposit vast amounts of silt in the tidal estuaries from Anchorage to San Diego.

The interaction that takes place when the sea meets the river causes the load to be dumped right at the mouth, but it is then sorted and distributed by the wave action and the currents. The interesting thing about the western coastline of the North American continent is that there is a 'long-shore drift' taking place from the Gulf of Alaska to the Baja Peninsula. This happens for a number of reasons, but mainly because of the configuration of the coast and the direction of the prevailing ocean currents.

The Alaskan current spills down through the Bering Sea and swashes silt from the estuaries of the Yukon south until it meets up with silt from the Athabasca. It continues like this down along the coast until it reaches the tip of the Baja Peninsula and there off-shore, it enters a submerged subduction zone, is sucked down into the molten magma chamber, and everything starts all over again.

Some harbors are superbly scenic and bright because of their individual formations and just a delight to sail into early in the morning or at sunset. Rias are drowned river valleys resulting from post glacial sea level rise, that get deeper gradually toward the center. Fjords are drowned glacial valleys and they are steep sided and deep, since they are essentially u-shaped and formed by the ice 'bull-dozer.' Fjords have rock island barriers, sometimes known as skerries, that block the mouth of the harbor and ships have to carefully navigate through the narrow exit channels.

Most large rivers have deltas where they drop their load upon reaching the sea. A delta is an area of mud flats and silt deposit, through which distributaries intertwine and wind to take the river out to the ocean. A delta changes shape and formation as the channels shift during flood and storm conditions, and can pose many problems for shipping. Often times, an aerial view of a delta will resemble the print of a large bird and they are called, aptly enough - *bird's foot deltas*.

The ocean and the seas act like giant radiators for our planet. Think about the engine in your car. How long would your car run if your radiator was broken or empty? The radiator distributes the heat from the engine and helps cool it down. In a similar way, the ocean acts as a giant radiator for the planet earth and helps distribute the temperature so that it doesn't overheat.

The water at the equator is constantly being heated by the sun, until you might think that it should boil. Meanwhile, the water at the poles is freezing so that we have thousands of feet of ice. And it would continue to grow, but for one critical law. Lucky for us, since it would make for a very strange planet if there were no inter-exchange of temperature. In fact, the mixing of cold and hot water is similar to the mixing of warm and cold air and it explains most of the reasons for our climates.

Water changes temperature more slowly than air and land do. Thus it takes much longer for the sun to warm up the ocean than the land. Oceans control the earth's climate because of their immensity, and the slow rate at which water changes temperature. The ocean has a modifying effect on land temperatures. It helps keep the air from becoming too hot or too cold.

The ocean also provides rainfall. Water evaporates from its surface because of the sun's heat, it rises and forms clouds. Winds carry the clouds across the land. And wind is moving air, caused by differences in pressure as a result of differences in temperature. This happens because of insolation, and unequal heating of the surface by the sun. Clouds provide rain and snow that feed rivers and lakes and help plants grow and sustain life. It all fits together wonderfully, to make the great picture!

Without the ocean, earth's climate would be like that on Mars - extremely cold at night and searing hot during the day. A climate of this nature would not sustain life as we know it and we can thank our stars today for the influence of the oceans, in creating this climate where we can live and breathe in comfort.

Water rises when it is heated. Cold water takes its place from below. This simple motion sets up a series of world wide ocean currents that follow a definite rule of direction. Hot water flows away from the equator and cold water flows towards the equator. Added to that, are a number of factors that aid and augment the pattern of flow that the currents take. The first is the direction of the prevailing winds; the next is the deflection of the wind and the water due to the rotation of the planet and the third is the configuration and shape of the continents that force the currents to flow in certain paths.

In the north Atlantic, for instance, the water at the equator is warmed and it is set in motion. It moves away from the equator towards the north, but because of the rotation of our planet it is deflected towards the left and enters the Gulf of Mexico. Here it is called the *Gulf Stream* and it circulates in the gulf creating a phenomenon called the Saragossa Sea near the Bermuda Triangle where sea-weed collects.

From here it drifts into the North Atlantic where it is called, oddly enough, the *North Atlantic Drift* and it continues north until it reaches the coast of Norway and the Arctic Circle. The warm water has a major influence on the climates and the living conditions of the people with which it comes into contact. The gulf is plagued with tropical storms and hurricanes while the west coast of Europe has a wonderfully, mild maritime climate, and the port of Narvik in Norway, which is inside the Arctic Circle, remains ice free and accessible all year round. Meanwhile, harbors in the Gulf of Bothnia are ice-bound for up to six months each year.

But the Gulf of Bothnia is ten or fifteen degrees nearer to the equator than Narvik, and should be warmer. However, warm water from the equator via the Gulf, helps to maintain the temperatures above freezing point even inside the Arctic Circle. This has momentous economic implications for places like Narvik, that can remain in operation throughout the year and can even ship goods from ice-bound Sweden to the rest of the world. Take a quick peak at northern Norway and locate the frozen waterways of the Gulf of Bothnia on your atlas to get a thorough understanding of economic geography at work on a localized scale. You begin to grasp how physical, environmental, political and economic issues intertwine to create an influence on how man copes with existence.

The oceans are our greatest natural resource, making life on this planet possible, by regulating temperatures and replenishing our water supplies. It is good to study oceans from the vantage point of the big picture paradigm, and see them as the nurturing force of the earth upon which many island continents float.

When we understand how the oceans interact with other geographical factors, like insolation, wind force, and the configuration of land masses to create the climatic conditions that prevail over much of the earth, we begin to see the importance of their presence for sustaining life on our planet. I am especially pleased for you that you can learn about oceans, because now when you go sailing you will have more to enjoy; more than the smell of the fresh sea breeze, more than the sun dripping onto the early morning water, and more than the eternal sounds as the seas flow and ebb in never-ending motion.

Chapter Eighteen:

What Price to Pay?

We have come a long way since the days when famine, plague and pestilence were our natural adversaries, and war our inevitable outcome - or have we? We may sit in the comfort of our heated homes, drive to contemporary offices in sophisticated automatic cars, and sway to the tune of the latest beat, but have we escaped the grip of nature and the struggle that has always been inherent in man's existence?

The question boils down to the dichotomous concern respecting nature and man's relationship to it. Nature is at once the raw material for the new-found freedom in our modernity. But in our blind and incessant quest for exemption from the harsh exigencies of necessity, we have managed to lay waste and pollute our very provider - nature itself. It's as if we were given all this technology and prosperity, but there was a hidden price to pay - a price that is only now dawning upon many of us. Look at Nobel. Today his name conjures up peaceful images, but remember what his name first stood for? Explosives! What was his price? Einstein, at one time, urged the United States Government to develop the atom bomb, but his views radically changed. What was his price?

What are your views on the fundamental questions that now plague our time? We cannot rely on industry and government to choose the pathway for us, and that is the prerogative of the democratic nation. The government must serve the people. You must be informed and aware of the issues and have resolved, for yourself, the answers. You must know that it is futile to try to stand idly by and hope that someone else will supply the solutions to these burning controversies.

Industry and government have too much at stake, too much vested already in their present courses of actions, relating to policies and short term futures. Do not expect them to make decisions that are likely to be to your liking. That is, of course, unless you let them know what you want and where you stand. Mistakes have been made in the past in all countries and we should be aware of the mistakes so that we can learn from them and proceed ahead with caution.

Recent riots in many countries and particularly in Great Britain highlight the blind 'tunnel-vision' policies that have been ushered in for years, without regard to the growing numbers of youth unemployment and their lack of involvement in decision-making procedures. How close are we in this country to a similar situation? Politicians might like to sweep these incidents under the carpet, and hope they go away, but they are very real issues in a global economy where technology is competing with human beings.

Many of our young people are already resentful and frustrated with the way things are and the seemingly futile mess that our generation has made of life. Suddenly everybody is asking why? Why would they riot? What are they trying to say? Why are they 'joy-riding' in stolen BMWs and looting department stores? Where is the logic in this kind of behavior?

All this mass confusion and anxiety points to the fact that modern man has arrived at a point in his existence where life has become meaningless for him. He has forgotten the meaning for his existence. Life has come down to a crazed quest for security and comfort but then, lost in his self-made quagmire which we call progress, man has no notion what to do. Progress culminates in the tragic recognition that life is meaningless. Maybe that was what Thoreau was describing during his time at Walden.

Technology is a double-edged sword; it has been our friend and our enemy. On the one hand it liberated us from starvation, death and misery, thus raising our standards of living and decreasing the mortality rate. But along with these wonders came an excess of headaches and irritations that will not go away, each one powerful enough on its own to bring destruction to our life on the planet. We have inherited a new set of problems that we have to recognize, understand and control so that our very existence is guaranteed. That is where you come into the picture. You must be the person to understand and control these issues for yourself, so that you are a responsible citizen in today's world. We all need to know what is going on and we have to make informed decisions whenever we get the opportunity to do so.

We teach our children to exploit nature for the good of man on the one hand, and yet we try to implant a solid reverence for our planet at the same time. Is this possible? Can we serve both masters? What are we saying to our children - that it is OK to exploit as long as we achieve progress. Wait a minute? Is this right? Can leaders of industry meet so-called financial objectives, raise general living standards, wages, and conditions, and still bear in mind that their activities often result in negative and harmful impact on our environment.

I guess the fundamental question surely is this: is it possible to have a sound planetary vision in our modern world? Can you, in the pursuit of whatever you do on a daily basis, do anything different to help get the message across? Where do you stand on all these issues? These are fundamental questions that need to be asked and answered by each and every one of us today to insure our tomorrows.

What is the planetary vision of modernity?

There is a conflict in the modern world between the exploitation of nature and the preservation of our pristine resources, culminating in several themes that describe impending disaster in our world. These are serious issues and be aware that they can impact you and your children. They include the phenomenon of population explosion, all kinds of atmospheric pollution, the imminent threat to our water supply, the growing need for urgent waste management, the ruinous surge in species extinction, destruction in our forests and depletion of our agricultural lands by encroaching deserts. Each one of these issues is, in itself, a threat to the very existence of life, as we know it, on the planet.

Look around you. The cities are too large and vastly overcrowded, the free-ways are chock-a-block, and the national parks are overflowing with visitors. Yet we live in a part of the world where there are controls on population and we are not headed for disastrous overpopulation predicaments in the next decade or so. Meanwhile countries in what we call the third world are experiencing massive increases in population for a number of reasons, each related to the growth and progress of the developed countries. Prominent among them are the changes in life expectancy brought about as a result of progress in medicine. In addition, improvements in agricultural production techniques have had profound impact on the living standards of poorer countries. Aid from concerned developed countries has brought immense change, though not always the desired and planned one.

How many worlds are there, anyway? The first world is sometimes referred to as the western world and includes the developed and advantaged places like the United states and Europe. The second world is the world of the communist block and China, and in recent years, is changing so rapidly that it seems to be merging with the first world. The third world refers to the underdeveloped regions of the world and is a major segment of our planet that requires serious maintenance. And now there is the fourth world, members of OPEC or oil producing countries that are rooted in the third world but have wealth and living standards more in keeping with the developed world. It conjures up images of sheiks driving Mercedes through the unpaved deserts.

Ever since Malthus, population questions have been a worry to planners and thinkers. Who or what was Malthus? Thomas Robert Malthus was a British economist who predicted certain troubling facts about the carrying capacity of our planet, around the year 1798. Malthus' hypothesis resided around the observation that population tended to grow in a geometric ratio, while food supply grew arithmetically. To those of us who are not mathematical wizards, what this means is simply, that population increases geometrically as - 2, 4, 8, 16, 32 ....., but food production increases arithmetically like this - 2, 4, 6, 8, 10..... or put in another way, population grows faster than food supply. When this occurs, Malthusian Checks were predicted to come into play to stabilize the process. These checks were none other than your everyday, famine, war, pestilence or other kind of human misery. Malthus was a cheery old chap.

This was in the Britain of eighteen hundred and, of course, Malthus was wrong. Science came to the rescue of the miserable European peasant of the nineteenth century, accounting for great improvement in agricultural techniques, the opening up of the new world, major advances in medicine and improved living standards in general. All this added up to a more stable environment that resulted from an increase in food supplies.

Again in the twentieth century, after the second world war, it looked like Malthus' predictions would again ring true. But, for a second time, science again came to the rescue and the remarkable progress made in food production, in what has come to be known as the *Green Revolution*, made major impact in deterring a serious world famine. But we are not out of the woods yet. Look at your world today and pick out the places where famine, war and pestilence is a way of life for millions of wretched inhabitants who go about their daily plight blissfully unaware of either Malthus or you. Many countries today are experiencing staggering population growth, lending credence to the prediction that the earth's population will be too great for the food supply by the middle of the next century. There's an interesting statistic to leave with your grandchildren, when you go.

Another impending disaster resides about the frightful condition of the air that we breathe. On top of that, our atmosphere is being modified so that harmful ultraviolet radiation from the sun is not being filtered out like it should be. Air pollution and ozone depletion are common terms that many people volley around with little comprehension as to what is going on or how it impacts them. Simply stated, where man brings industry, pollution is soon to follow. This lesson was first learned in the predominantly 'dirty' industries of the industrial revolution in Great Britain and Germany in the last century. Coal was a major contributor to the toxification of the breathing air, as factory chimneys pumped thousands of tons of grime daily into the atmosphere without any regulation.

Today there is an added source of destruction to our atmosphere in the form of carbon monoxide from exhaust fumes of cars and trucks that proliferate in the most developed countries. It seems that the more developed we are the more damage we can cause to the environment. What injury does the forager in the Kalahari desert do to his environment in comparison with you, as you go about your daily tasks. Paper or plastic?

The more developed we are the more damage we can cause to the environment.

Smog and man-made chlorofluorocarbons resulting in a hole in the ozone layer could be the most serious threat to our future since the nuclear bomb. Now, there's another pleasant thought to leave to your grandchildren. Hey, we're modern man, we are the developed countries and we know what we are doing.

Take a minute to count the number of nuclear spills and forced clean-ups that have taken place by accident in your lifetime and you will begin to see the enormity and the gravity of the situation. But we won't be around to deal with it - so let's plough on regardless.

What about acid rain? Do you understand what it is and what it does? When rain falls through the carbon monoxide gases it picks up some of the carbon oxides and becomes dilute carbonic acid. In other words you could say on the next rainy day that it is 'carbonic aciding' outside right now. Would you want to leave your car out in this kind of corrosion? But we do all the time. Look what is happening to the forests and fisheries in neighborhoods adjacent to the major industrial locales.

Third among my list of serious conditions on our planet right now is water management. Water is probably the most essential ingredient for mankind and we tend to take it for granted as we use and waste millions of gallons daily. It is only when we are faced with a shortage of water, for whatever reason, that we suddenly are reminded how important to us it is. And yet we treat it with disdain and contempt on a regular basis.

There has always been the same amount of water on the planet. It is used over and over again as it runs through the natural hydrological cycle. But today, for the first time there are very serious doubts about the carrying capacity of the water supply for vast tracts of agricultural land and for entire cities here in the United States. The reason for this devastating state of affairs is none other than man's mismanagement of a once prolific natural resource.

Let's think about that statement for a minute and put it into context with some other events that have taken place in our lifetime. Oil, coal, peat, all forms of fossil fuels took millions of years to grow and proliferate as each was laid down in sedimentary formation, layer upon layer through time. And yet in the one hundred or so years that man has found a way to utilize these fuels to improve his condition, these fossil fuels have been depleted to the point of near extinction. Meanwhile, the disastrous ill-effects of using these fuel sources has been felt worldwide and will continue to impact each future generation for a long time.

Has man learned from these mistakes? Were they even mistakes? It depends on whom you ask. Where do you stand on these issues? What will we do when all the coal is burned and all the oil is used up? The new forces of energy are so complex and so powerful that we have within our hands the very means to end life as we know it on the planet. On the other hand, many scientists have spent their lives experimenting with alternative sources of energy, sources that have nothing to do with fossil and nuclear fuels. We are now familiar with the issues surrounding solar power, wind power and water power. It always seems to boil down to economics, however. Answer this question for yourself. If there was no more oil remaining in the world; if you got the very last gallon of gas, do you think that there would be nothing to replace it? Would we all just sit around talking about it? Or would we immediately have a replacement system for it, one that is developed today, but since there are still huge oil reserves stockpiled in many countries and since there is already so much vested on the status quo, no innovation is economically feasible.

Toxic wastes and nuclear fall out. Can you believe this? We are talking here about the very real dangers of contamination from hazardous radiation as a result of man's ability to control nature. This was what he strove to accomplish for hundreds of years. The ability to be in control of his lot. And now look at where he is. Who is in control? Who will even tell us?

Things happen with such incredible velocity and severity that maybe we don't even want to know. But that is a worst case scenario and we cannot go around living our lives with the constant fear that the next breath we draw might be our last - because that's right, it very well could be our last. So what? There is nothing we can do about it - or is there? Let's stand back and see what we can do today to help survive in the world of the twenty-first century.

We produce massive amounts of waste material in our modern world, in fact millions of tons each year. The bad news is that the production is increasing rapidly. It seems that the more progressive we get the more waste we create. There are many kinds of waste; gaseous waste in the form of carbon monoxide from cars, liquid waste such as sewage, and solid waste including paper and plastic, glass bottles, aluminum and steel cans, garbage and junked automobiles. Refuse of this nature looks ugly, smells foul, and attracts rats, insects and other vermin that spread diseases. And we produce it, and produce it, and ....

Not only are we producing more waste, but we are also producing more and more wastes that are increasingly more difficult to control and dispose of safely. The price of using the latest technology appears to be the spawning of a whole new regime of hazardous and toxic materials that are not going to go away quietly. We grew up in the fifties, sixties and seventies with the notion that things were bio-degradable and would rot back into the system with little harm to anything or anybody, but the message of the nineties is far from bio-degradable. Today, we need to know about the half-life of elements and the resilience of materials to decay. Tin cans, for example, that used to rust and become part of the soil are being replaced by aluminium cans that stay in their original state for years. Paper packaging that decays and burns easily is being replaced by plastic packaging that decays slowly and emits dangerous gases when it is burned.

> **The price of technology appears to be the spawning of a whole new regime of hazardous and toxic materials that are not going to go away quietly.**

People dirty the air with gases and smoke, poison the water with chemicals and other substances, and damage the soil with too many fertilizers and pesticides. Air, water and soil are the essential ingredients for sustained life on this planet - necessary for our very survival. Badly polluted air can cause illness and death. Polluted water kills fish and marine life, and polluted soil reduces the amount of land available for growing food. Environmental pollution brings ugliness and death to the naturally beautiful world. Sewage flows into lakes, oceans, rivers and streams sometimes as treated sewage called effluent, but in many cases, as raw sewage that is foul-smelling and contains disease-producing bacteria. The more people, the more sewage; since it originates in the sinks and toilets of our homes, restaurants, office buildings and factories. It is not a pretty picture.

Pollution is one of the critical examples of man's head-on collision with nature and the abuse of nature. On the one hand we can control nature with our progressive technology, but on the other hand, we destroy the very hand that feeds us, by polluting, poisoning and choking the natural environment. It comes down to a case of basic common sense. Do we want to carry on creating bigger and better freeways, larger automobiles, more and cheaper food or do we want to plan for the future of our planet and guarantee the survival of our children? Look what we have prepared for our children already.

The megalopolis in the eastern United States has been polluted beyond belief because of the sheer numbers of inhabitants utilizing the natural corridor. A good place to raise your kids! The Mediterranean Sea is the recipient of numerous major rivers that deposit effluent into the natural reservoir. Would you want to swim, here? Nearly every major city has felt the disastrous effects of waste disposal problems relating to refuse, solid waste and air pollution in the last twenty years. Some cities, like London, have had to ban fires and have spent millions of pounds to clean up their rivers and buildings. Great jubilation was reported recently at the sighting of salmon in the Thames after so many years of sludge and gunge. What price progress?

Fish from certain parts of the world cannot be eaten for fear of mercury poisoning and other toxic diseases. Crops from certain farms cannot be eaten because of abuse of fertilizers and pesticides, and yet we continue to develop and use these ingredients despite the fact that they leach into the soil and end up in run-off, and into the rivers and eventually the seas. Try this little exercise to really get a good picture of the destruction and havoc we have wreaked on our planet during our short life time. Draft a map for your children, one that they can take with them into the future. Call it: A Map for My Children.

"Places to Avoid; Where NOT to Go on Our Planet"

Produce this plan for the next generation to show our children where they should not go on the planet for fear of their lives. This should be a list of all the places where you would not live right now because you either know that there is foul air there, contamination, shortages that could lead to famine, sickness, disease and so on. This little project would be quite the eye-opener for the politicians and industrialists who still do not admit that we have to stand up and be counted for our planet.

Bikini Island is still a no-go area since numerous nuclear tests resulted in radiation contamination. You don't want to live in Tokyo because there is vast over-crowding and chronic air pollution. Mexico City boasts the worst urban environment on the planet. Regions near toxic industrial plants, like Sudbury Ontario and the Soviet Kola Peninsula are barren waste-lands, devoid of vegetation. The once pristine American west is now a sullied and dirtied extension of our cities; a result of power plant emissions and motor vehicle exhaust. Parts of Brazil are so polluted that they should be designated disaster zones. In addition, there is an estimated two hundred thousand square miles of forests in industrial countries that have been adversely affected or destroyed by acid rain and other toxic forms of air pollutants. Many lakes, including the Great Lakes, and inland seas like the Caspian, the Baltic and the Mediterranean suffer from large coastal populations and industry, resulting in pollutions and poisons beyond belief. Ports like Jakarta, Bangkok, and Manila brew noxious stews and nauseating aromas.

Remember PCBs? There were high concentrations of PCBs in San Diego Bay; polychlorinated biphenyl that does not break down like other organisms and they accumulate and remain in the environment for many years. Fish are contaminated by them and then humans can be adversely affected, causing cancer and birth defects. Remember DDT? At the time this ingredient was thought to be a life-saver. It was used liberally to clean up the lice infested conditions during the war, and was a household mainstay for many years. But now we realize the damage it did to the natural food chain, the contribution it made to the extinction and near extinction of certain species and how it effects humans.

And the problems are only getting worse. Cities like Seattle, traditionally pristine and once rated the most livable city in the United States, now boasts a shortage of water because of the pollution of the natural water-ways by agricultural run-off, pesticides, and industrial spills. Look at the catastrophic oil spills that occurred in the Pacific Northwest and in Alaska, over the last few years. How often have we seen politicians and officials run around pointing the finger at the other person and tying resources up in bureaucratic red tape, while thousands of gallons of crude oil spill into the seas off our coasts. Beaches in Alaska, British Columbia, Washington, Oregon, California, Massachusetts, New York, New Jersey, Chesapeake Bay, Florida, and the Gulf Coast were ruined by oil spills in the eighties and nineties. But wait, that is just about all our coast-line! And we are not alone in this respect. Beautiful stretches of coastline in France, Britain, Ireland, Spain and places as antipodean as Norway and Australia, were also destroyed by the drive for progress and modernity. In truth, no place is safe from the ravages of industrial waste.

With the advance of the nuclear age, the issues regarding waste are even more critical since the potential for destruction is so much greater. There are many forms of waste from the nuclear plants; thermal water pollution that kills fish and damages the marine ecology, radiation waste and leaks that are life endangering and cause for serious debate as to the feasibility of such plants. Nowhere is the dilemma of man's relationship with nature more real than in the nuclear debate. Nuclear energy promises to rid man's dependence on the traditional fossil fuel as a supply of energy and it also offers hope to provide unlimited energy reserves for the future. But in addition to all this glitz, comes the very real danger of contamination and total destruction; that's the worst scenario. A gigantic waste disposal headache that involves our children's grandchildren constitutes the warm and fuzzy scenario.

Unlike fossil-fuel plants, nuclear plants do not release solid or chemical pollutants into the atmosphere. But as the number of plants grows so also does the possibility of serious disaster. Such has already been the case at Three Mile Island in Pennsylvania and in Chernobyl and the Kola Peninsula in the Soviet Union. Another and long-term fall-out from the nuclear debate is the question of what to do with the waste product that results from the fission and fusion processes. There is no point in saying that we do not need this technology - because clearly we need something. But what to do with the nuclear waste? Now there's a dilemma! Damned if you do, and damned if you don't.

The problem lies in the fact that the wastes remain extremely radioactive for thousands of years, a result of the presence of plutonium. These elements can cause cancer and genetic defects in people. And as yet, no permanent resting place has been designated for the residues. So where are they, right now? They are presently cooling, temporarily, on the grounds of the nuclear plants. Experts cannot agree on what storage facility will be the least hazardous for our great grandchildren, yours and mine. Yes, our grand kids will need special instructions relating to the whereabouts and potential danger of the final waste site, in case they disturb it so that radioactivity is accidently or deliberately released. What a nightmare!

The paradox is people. Without people there would be no pollution, since nature would take care of itself. So now we introduce people, but we must instruct them how to take care of things, like nature would. It is a matter of balance. We are all guilty of craving the need for security and comfort, but we rarely think of the price. Consumerism, greed and short sighted policies by each of us and particularly by governments only lead to wanton destruction of our natural beauty.

Man causes pollution.

Isn't it interesting how most pollution is caused by things that benefit people. For example, exhaust from automobiles causes a large percentage of all air pollution. But automobiles are part of our way of life, our very culture. They provide transportation for millions of people on a daily basis. Mobility is one of the keys to the modern way of life. Similarly, factories discharge pollutants into the air and waterways, but factories provide jobs for people and produce goods that people want. After all, we created factories and the entire work ethic surrounding that method of production. In order to produce more and cost-effective agricultural produce, scientific applications have been used more and more in recent years. Too much fertilizer and pesticides can ruin soil, but fertilizers and pesticides are important aids for crop production. And so, we are again stretched on the horns of a dilemma, a man-made dilemma, mind you, but still a dilemma.

Man can cure pollution.

What can you do to change this picture of world destruction in your neighborhood? Yes, in your own house, because it is going to be won right there in each person's backyard? For a start, smaller, more efficient cars are in vogue thankfully. There is even a drive towards electric cars and solar technology; in other words, alternative energy. Alternative, that is, to fossil fuels from whence the carbons in our air originate; and this is good.

Technology may offer another method of combatting over-crowded cities and air pollution by creating the work environment at home. Today, by utilizing telephone lines, modems and fax machines, in conjunction with your PC, it is possible to be just as, or even more, productive at home. This way you can avoid entering the traffic fray each morning and evening in the struggle to get to and from the office. I have been involved in a work related experiment over the last three years, using remote locations like this for regular employees.

It all began by accident one bad winter's day when snow prevented most of my staff from reaching the office. Those who had modems carried on with business as usual, and were able to report that they got more accomplished because there were no meetings to interrupt them. Everybody seemed to be getting modems after that. They enjoyed working at home and they liked having more control over their own hours. On top of that, the onus was off me to manage them and each person became responsible for his/her own schedule. We were able to leave the freeways open for the millions of commuters who read about this kind of thing and sit in their carbon monoxide-producing traffic jams, day after miserable day thinking about us. I'm not saying that telephones and modems, and fax machines can replace the office structure entirely, but it is worth thinking about as an alternative for what we've got on the ground today.

Recycling is a growing phenomenon in today's society and has resulted from the flagrant misuse of the environment by so many people for such a long time. There has been a growing increase in recycling techniques among enlightened environmentalists since the sixties because of concern over the dwindling supply of natural resources and the increase in environmental pollution. Recycling is now big business, but it is doing us a all a great favor. Cans, paper and glass can be recycled and used over and over again, conserving our natural resources and saving our natural beauty.

Progress has a price, there is no getting away from that fact. Sometimes the price is unacceptable but there are other ways. Many people have overcome the megalopolis dilemma by changing their life-styles and making conscious decisions not to live 'like that' anymore. Some end up moving to the mid-west to some small town where they spend their time in relative tranquility and fulfilled pleasure. They may not be in the fast lane, but that was what they were trying to avoid with their move. They place more value on fresh air and home-grown vegetables than access to the international airport or the health club. These are life-style decisions that people deliberately make because they have experienced enough of the hustle and trauma of life in the over-crowded city. They reached their limit; too-much looking for a parking spot, too-much fighting the elevator, too-much jostling with crazy people who were just like themselves, and then finally said - enough.

Enough.
Enough.
I've had enough!

And then what. Hide away in the solitude of the small country town, trying to eke out an existence in another community doing something that pays merely a pittance. For most of us this is not really a suitable option, and certainly not the panacea for the frustrations and ills of modern life. After all, we have children to raise and we need schools and we have to support them, feed them and clothe them. And anyway, the kids wouldn't live in the boondocks. Is there any place left anymore where nature is pristine and man has not injured it?

Many everyday activities harm the delicate balance that holds our natural fabric together, including, farming, hunting, logging, settlement and growth of urban areas. In many countries, especially in third world areas, poor farming practices have a devastating effect on wildlife and natural habitat. Meanwhile, in the more developed countries the spread of cities and creeping urbanization is eating up vast amounts of previously open country. Pollution poisons air, water, plants and animals. Pesticides and industrial wastes pose a serious threat to many species. While extinction has always happened by natural selection and for other reasons, relating to climatic and environmental changes, extinction has occurred at such a rapid rate in the last two hundred years, that it is cause for heightened alarm. Over fifty species of birds, seventy five species of mammals, and hundreds of other animals and plants have become extinct in recent history, or since man's increased prowess with science and new knowledge that helped him survive. Human activities and population growth have increased the danger for our natural wildlife, not just here in America but in all parts of the world. As a result, some species have declined greatly in numbers and others are now extinct. This highlights and intensifies the need for conservation.

At first, the development of increasingly efficient weapons had a direct impact on the indigenous wildlife as man hunted and killed vast numbers. They may have started out hunting different kinds of animals and birds for necessity, food and clothing, but as technology improved, they were able to equate profit and better living conditions with the increased production of pelts, or ivory or mink and so on. This soon lead to the endangerment of certain species and the annihilation of others. Sad as it may be, it happened and can still occur again, any time we let our guard down and look at life through jaundiced spectacles.

But this was not the only way that we managed to get rid of some of our wildlife species. Since man first discovered the techniques of agriculture there has been a desire to improve the yield, increase the output and make the whole operation less labor intensive. Technology and science was a perfect tool for taking the hardship out of farming, while at the same time feeding the teeming millions that needed food. But at what price? We cleared forests, drained swamps, and dammed rivers, in our efforts to open up more farmland. In addition, we increased the size of fields by getting rid of ditches, felling trees and excavating rough pasture land, to make way for agricultural land. As the population grew and man prospered, more industrial buildings, homes, and roads were needed to provide the appropriate infrastructure for an industrialized society. The combined end result was the eviction of the native animals and birds to remoter habitats, some to endangerment and others to extinction.

About two hundred and ten kinds of birds throughout the world have become so rare that they are considered to be endangered; sixty-five alone in the US, including the Bald Eagle, Californian Condor, and the Whooping Crane. Hunters and trappers killed many Bald Eagles and Ospreys. Pesticides, especially DDT contaminated the food supply of the Bald Eagle and the Whooping Crane. One result was that the eggs laid by the birds were so delicate, with such thin shells that they cracked under the weight of the incubating parent. Illegal hunting, illegal egg collecting and loss of habitat all contributed to the decline of the Californian Condor.

We are the losers when a species becomes extinct.

If we ignore the need for conservation, today's endangered species will become extinct and many new ones will face extinction. Several hundred species of animals and thousands of species of plants still face that danger of extinction. We all lose when this happens; we lose something of intrinsic value that is beyond material price, that can never be replaced. Our lives are impoverished beyond calculation, and our future generations can only speak of it, like we do now of species like the Dodo and the Auk. Every species plays a vital role in helping maintain the balanced living systems of the earth. These systems must continue to function if life is to survive on our planet. The loss of any species threatens the survival of all life. And if we stop and take the time to think, it even threatens our own survival.

Other natural disasters and man-assisted destruction also threaten our time on this planet. Alarming increases in desertification have occurred in the past few decades, caused by prolonged drought, coinciding with population growth, over-grazing, and the closing of political boundaries to disrupt the practice of nomadic life. The result is pressure on marginal lands generating the spread of desert conditions.

One seventh of the earth's land area is desert or semi-desert. The Sahara desert is four-million square miles. The natural forces that create deserts have not changed much for thousands of years. However, various human activities have caused desert regions to expand considerably. This expansion occurs because of the continual loss of fertile land on the outskirts of such regions. Salinization has strangled irrigated fields in Australia. Irrigation was supposed to be the major scientific break-through to rescue desert lands but evaporation and capillary activity resulted in the poisoning of the soil by salts. People destroy millions of acres of this land yearly. Over-grazing of livestock is the chief cause of the loss. Other causes include mining, improper farming methods and destruction of trees.

We can take steps to curb desertification, including plant-ing of trees to reduce the wind at ground level. This revegetates desert regions and prevents sand being blown onto crops. Mass planting of special root crops and grasses helps reclamation and stops the spread of deserts. By improving farming methods the land is not impoverished to the point of abandonment, and the future of the soil is cared for. But the greatest achievements can be attained by limiting the number of livestock in desert areas, thereby preventing overgrazing and defoliation. Even as we think about preserving our agricultural land from detrimental encroachment by deserts, we have to balance our view by ques-tioning the legitimacy of opening up new lands at the expense of our natural forests. This is a complicated and serious question that needs prompt attention.

One of the greatest catastrophes that is staring squarely at us is the global deforestation that is occurring even as we speak. Forests may seem a long way off, when we sit in our offices, resplendent with beautiful antique globe on our executive desks. Out of sight, out of mind. But when the day falls that we are advised to stay indoors because the air is not fit to breathe, we look to the source of our oxygen with sad drawn faces and wonder how we could have allowed this situation to occur. But it is not me, it is not you. These forests are being depleted in the southern hemisphere, they are being felled and burned in South America in the Amazon Basin, where ever that is.

Stop.
Stop.
Stop.

Our forests, yours and mine, are our last greatest store-houses, where everything is rich, lush and abundant. We must know about them and recognize what they meant to each of us, in the past, and especially going into the future. Tropical forests shrink annually by an estimated eighty-thousand of their total ten-million square miles. Say that slowly; eighty-thousand square miles of forest disappear every year. This occurs for the very same reasons that animals and birds are becoming endangered or extinct. The culprit is once again, man, as he prospers and becomes more comfortable and secure in his new-found scientific freedom. These reasons include expanding agriculture, logging, development, demand for fuel, erosion, landslides and floods.

At one time about sixty percent of the earth was covered with forests. Today, it is less than thirty percent. I shudder to think of the future, and what figure we could attach to it. Forests have always been a source of great importance for human beings - hence the need for their continued careful management. Forest resources unlike other fossil fuels are renewable, and help conserve and enrich the environment in several ways. First, they soak up large amounts of rainfall, preventing rapid runoff that causes erosion, gullies and flooding. In addition, rain is filtered as it pours through the soil and becomes ground water providing a clean fresh source of water for streams, lakes and wells.

Forest plants like all green plants help renew the atmosphere. As the trees and other green plants make food they give off oxygen and remove carbon dioxide from the air. People require oxygen. If green plants did not continuously supply new oxygen all life would soon come to an abrupt halt. A comforting thought. And if carbon dioxide built up in the atmosphere, it could severely alter the earth's climate.

Forests provide a home for many plants and animals that can live nowhere else. Without the forest, many kinds of wildlife could not exist. And joyfully, forests are a source of pleasure to man, where we can delight in camping, hiking and thinking in the midst of enriching scenery. Here is the final place to savor relaxation in quiet and beautiful surroundings.

So where does all that leave us? We are at the delicate crossroads of technology and nature. Should we look to more use of technology, in our search for a cure to the ills of society, to counter-act the bad impacts of our use of that technology. What a vicious.circle. Is that the way it is meant to work? Technology, creates a problem, while solving a problem, so we look to technology to fix that problem and it goes on...

Technology - a solution?
Technology - the cause?

The problem is that, many people will go as far as they can, to milk the system and take from nature, as long as they think it won't impact them, personally. Bad mistake! We are all part of the global village today. It may happen on a different hemisphere, but yes, it does impact you. An oil well burns in Saudi Arabia, and your sunset is spectacular next week. A factory emits powerful carbon fumes in Canada, and your rainfall is acidic for months. Animals and fish are contaminated with pesticides and drugs in Japan or South America, and our food chain is interfered with for a generation.

We cannot ignore what is going on around us in the world and say that it doesn't impact us. Where do you get your food supply? Where does your water come from? What is falling out of the sky on you every night? What rays are penetrating the earth's protective layers each day? What other unseen rays and unheard sounds are emanating from technological innovations in the work place? It is all very complicated nowadays. But the one sure thing is that one cannot drift along hoping for the best. You must be prepared and you must know how to find out what to do, and when to do it.

We must be ever-vigilant and we must let the leaders of our future know that we are the watch-dogs of the planet, not only for ourselves but for all our children. We have to stop looking back; we must not maintain those historic differences, that divide and separate races and lands; differences that allow awful disasters to occur.

In today's global village there is no far corner of the planet.

Instead we must look to the future from our new enlightened platform, as if it were out in space, where we can see the planetary ballet in full flight and realize that we have to take care of our home. Nobody else will do it for us. It comes down to you and me.

Epilogue

Bringing It All Back Home

The end is really the beginning - I learned that from James Joyce's book, *Finnegan's Wake*. Naturally, I didn't understand it, and in fairness, who could? His book commenced in the middle of a sentence, and when I finally reached the conclusion, I found myself back at the beginning. Very devious, I thought, because on the last line of the book were the first two words of the book. All this had something to do with Joyce's river theme, the river of life, with no beginning and no end. His fluvial analogy was designed to portray life as a constant revitalizing progression and a perennial flow through space and time.

Geography is a bit like that, with a limitless boundary that does not clearly define a beginning nor an end. In truth, you can jump in at any point, and drift along in the flow in any preferred direction, for as long as you wish. This book is a little like that also, and I made particular choices about direction and course. You might not agree with the decisions I made, and many of you may say that I omitted this piece or I never mentioned that area, and yes, you will be correct. There is no one-minute geography. Geography is such an all-encompassing discipline, that I could keep writing forever and still not cover all its intricate permutations. But, the decisions I made were carefully selected to comply with the basic cognitive premise that one should progress from the particular to the general; from the concrete to the abstract. I also took the liberty of assuming that we were all beginners, starting from scratch, with little or no previous learning in the geographic arena. And from there we set out; this is where we've been.

We began at the beginning, with the first tentative attempts to put some order and meaning to the events and phenomena of our universe. Some of these early attempts were ingenious and are still with us today, while others were utilitarian and served the period well, but did not withstand the test of time. Together we explored the definition that geography was a study of man in relation to his environment, and we investigated his eternal quest for solutions to age-old questions concerning our planet and our universe. In doing so, we discovered the many and varied scientific disciplines that contributed to the evolving picture of what we call today, the unified theory.

Against this colorful backdrop, we were introduced to the very building blocks of geography itself, the physical to-pography. In particular, we explored rivers, glaciation, oceans, deserts, vulcanicity, and the structure of our planet. This is the kernel of a good understanding of what makes up geography. It is also a perfect place to position yourself for the inevitable journey into the scientific world of the modern cartographer, the space-age town planner, the environmental controller, or the statistician. I chose not to dwell in depth on the quantitative revolution nor on the statistical analyses of geographical phe-nomena, but you should know that you can be ready to focus on any of these disciplines, given your solid preparation and growing understanding in your field of endeavor. And now it is time to put the pieces together and bring it all back home.

This is why I say, since we are at the end of the book, that we are really at the beginning of your geographic quest. This is but a stepping stone, a solid platform, from whence you can spring forth in your chosen direction. I cannot tell you which road to take or which branch at which to excel, and in fact, you may not know yourself, for quite a while. But that's OK. Give yourself time to explore this vast new field, and enjoy the process along the way.

For some it may happen overnight, but for the rest of us a gradual immersion in the elements and process of geography must take place before it all makes sense. As you begin to see more and understand the concepts and issues a little better, there will come a day when the 'readiness is all'. Then the deductive instinct will kick-in and throw its radiance over all the earlier apprenticeship. This is the true secret to geographical literacy. I know I said that there are no secrets and that anybody can become proficient in geography, but this formula applies to all our life. As you focus on something and learn more and more about it, there will come a time when the deductive instinct will reward you with the wonderful array of its riches.

There are recommended methods of applying this new knowledge and vision to your world, that will help you deal with the economic, political, and physical aspects of it. These plans are not mine, nor will they necessarily be your's, but they have been drawn up as guidelines for teaching geography by the Association of American Geographers and by the National Council for Geography Education. They provide a very useful framework for systematizing this expansive information and helping us understand the world in which we live. Geography is divided into five clearly defined themes that allow us put structure and order into the chaos that surrounds us. These themes are as follows: Location, Place, Human/Environment Interactions, Movement, and Regions.

The "Location" theme offers a starting point, asking the question; "Where is it?" Location, and in particular the application of map skills with regard to location, is fundamental to an understanding of geography. From there, we need to know more about "it". The "Place" theme posits the question; "What's it like"? This is where we can investigate the physical and cultural features of an area that give it a particular identity. How people respond to and modify their environment is also a central focus in geography and is the character of the theme; "Human/Environment Interactions". This encapsulates the very definition of geography itself, since it deals directly with man in relation to his environment. The "Movement" theme is a critical factor in bringing all the aspects of humanity and the physical environment together. This theme examines transportation and communication systems that link people and places. Finally, many areas are distinguishable by their unique characteristics and this is explored in the theme; "Regions". A study of the region helps us organize our knowledge about any land and its people.

This useful framework will help you organize your thoughts and lessons about geography, and with it you will make rapid progress. Use it to good effect, as you re-read this book. Incidentally, I know that there are hundreds of questions in here, and that each chapter probably prompts hundreds more. Most are answered directly, and others are answered indirectly, or the answers are inferred. But there is one that is not answered, and it is a serious question pertaining to your understanding of our planet and to geography. If you have come this far in the book you must have found this question. Have you attempted to decipher an answer to date? Try asking other people and see what they know, or think they know about geography? Good luck with your quest.

Me? I'm still sailing in the Puget Sound!

Index

A

B

H

N